OSPREY COMBAT AIRCRAFT • 5

BLENHEIM
SQUADRONS
OF WORLD WAR 2

SERIES EDITOR: TONY HOLMES

OSPREY COMBAT AIRCRAFT • 5

BLENHEIM
SQUADRONS
OF WORLD WAR 2

Jon Lake

OSPREY
AVIATION

Front cover
On three separate occasions during World War 2 RAF Blenheim pilots won Britain's supreme award for gallantry, the Victoria Cross, and many more Blenheim aircrew received DFCs and DFMs. You didn't have to be brave just to get into a Blenheim, despite jibes to that effect, but the aircraft's poor performance and frequently high-risk deployments did mean that conspicuous gallantry was often required to carry out some of the missions assigned. Blenheims were active during the years when Britain's back was 'against the wall', and great sacrifices were demanded in the name of national survival. Blenheims usually operated by day (mainly to show the world that the RAF *could* operate by day) and normally flew over enemy territory or enemy coastal waters without fighter escort or fighter cover. Accordingly, the aircraft suffered very high loss rates, and on numerous occasions only one or two aircraft from an entire squadron would limp home, the remaining bombers having been scythed from the sky by fighters or flak.

By the end of 1942 the Blenheim was judged to be too vulnerable to be used over Germany, and No 2 Group's squadrons re-equipped with new types. However, a handful of units received a new version of the Blenheim in the shape of the Mk V. Earmarked to support the Anglo-American invasion (Operation *Torch*) of Vichy French controlled North Africa, the Blenheim V quickly proved to be as vulnerable to fighters and flak in North Africa as earlier versions had been in Europe – losses were alarmingly heavy. On 4 December 1942 Wg Cdr Hugh Malcolm, OC No 18 Sqn, led ten Blenheim Vs from No 326 Wing in an attack against an enemy airfield near Chouigui, which Malcolm's formation (already down to nine aircraft) failed to find before they were attacked by an estimated 26 Bf 109Gs from I. and II./JG 27. Malcolm led his unit back towards friendly lines, hoping to thwart the enemy fighters by maintaining close formation, but one by one the Blenheims fell to the attacking Messerschmitt pilots. *Experte*

First published in Great Britain in 1998
by Osprey Publishing,
Michelin House, 81 Fulham Road,
London SW3 6RB

ISBN 1 85532 723 6

Edited by Tony Holmes
Page design by TT Designs, T & S Truscott
Cover Artwork by Iain Wyllie
Aircraft Profiles by Chris Davey
Figure Artwork by Mike Chappell
Scale Drawings by Mark Styling

Printed in Hong Kong

EDITOR'S NOTE
To make this best-selling series as authoritative as possible, the editor would be extremely interested in hearing from any individual who may have relevant photographs, documentation or first-hand experiences relating to the elite pilots, and their aircraft, of the various theatres of war. Any material used will be fully credited to its original source. Please write to Tony Holmes at 10 Prospect Road, Sevenoaks, Kent, TN13 3UA, Great Britain.

Oberleutnant Julius Meimberg downed three of what he thought were USAAF Bostons, six of the Blenheims falling close to the target. The three remaining aircraft conducted a fierce running fight as they ran for home, but were all eventually shot down, crashing or crash landing in British held territory. Malcolm's was the last Blenheim lost, falling in flames. Both he and his crew were killed in the crash, the wing commander later posthumously receiving the VC (*Cover artwork by Iain Wyllie*)

Title page spread
Seen flying over France in early 1940, this No 139 Sqn Blenheim IV (L8756) not only survived the carnage and slaughter of the Battle of France, but went on to serve with a succession of units, and was still active with No 9 Air Observers School and No 12 (Pilot's) Advanced Flying Unit as late as 1944 (*via Bruce Robertson*)

CONTENTS

GENESIS OF THE BLENHEIM

During the late 1930s, the RAF was rapidly re-equipped and expanded, with Bomber Command lying at the centre of that expansion. And the aircraft which formed the backbone of that expansion (and thus the aircraft which Bomber Command took to war in huge numbers in 1939) was the Bristol Blenheim, which has been condemned by history as inadequate.

Although the 1938 Munich Agreement had provided a useful breathing space during which Britain frantically bolstered her armed forces, the nation was singularly ill-prepared for war with Germany when it came on 3 September 1939. Nowhere were Britain's inadequacies more marked than in the air, and especially within Bomber Command. The new Wellington and Hampden medium bombers were effective enough (though they were soon found to be far too vulnerable to be risked in daylight raids), whilst the three new 'heavies' (Halifax, Manchester and Stirling) showed promise, although were not even close to service. But for tactical daylight bombing and close air support, the RAF relied upon unsuitable and entirely inadequate types. The Fairey Battle was poorly protected, poorly armed and pitifully slow, although it carried twice the bombload of the biplane Harts and Hinds it replaced twice as far.

By comparison with the Battle, the RAF's other light bomber (the Bristol Blenheim) looked very good indeed. The RAF's fastest bomber when it entered service, the Blenheim was said to be able to out-pace contemporary fighters. But 'faster than the fighters of the day' did not mean faster than the Spitfire or the Bf 109, nor even than the lumbering Bf 110. The Blenheim was faster (by a whisker) than obsolete biplane fighters like the Gladiator, or the Gauntlet. But even here the performance edge was not always what it seemed. While the Blenheim I could attain 260 mph, it

The original Bristol 142, christened *Britain First*, is seen here in its original configuration possibly taxiing out for its maiden flight on 12 April 1935. The aircraft soon gained a nose-mounted navigation light, and exchanged its four-bladed wooden propellors for three-bladed variable-pitch airscrews. It differed from the subsequent production Blenheim in having a low-set wing, and a less heavily glazed nose. Registered G-ADCZ, the Bristol 142 never carried its civil identity, instead receiving the B-class identity R-12 for testing at Martlesham Heath, and then the Air Ministry serial K7557. The aircraft served as a hack at Farnborough until 1942, when it became an instructional airframe

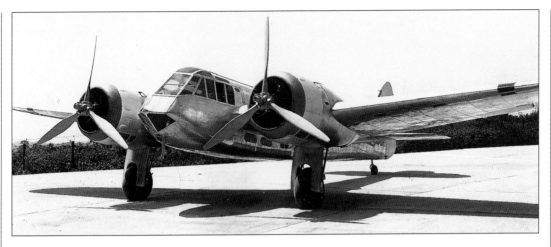

The production military Bristol 142M differed from the original 142 in having a mid-mounted wing and a raised tailplane. These changes were made to allow a shallow bomb bay to be fitted below the main spar. The Blenheim (as it was soon christened) also had a redesigned nose, with provision for a prone bomb-aiming position

cruised at only 200, and less when fully laden, carrying external bombs or trying to reach its maximum range. A 200-mph bomber was far from impressive (14 mph faster than the Hind it replaced!), and a 180 mph bomber (as the Blenheim often was) was quite simply inadequate.

The problem was primarily that the aircraft was too small and too modest, relying on feeble engines and carrying only a tiny bombload. The aircraft was too consciously a replacement for aircraft like the Hind, and the Hart's light-bomber concept did not translate when the replacement was a sophisticated monoplane bomber. Near-contemporary American light bombers like the Baltimore, Boston, Maryland and Ventura carried double the bombload and were powered by massive twin-row radial engines like the German Junker Ju 88 and Dornier Do 17. All were considerably heavier than the Blenheim and required longer, drier and smoother runways, but they were fast, well armoured and adequately defended.

If there was a place for an aircraft carrying only a 1000-lb bombload, then the Westland Whirlwind fighter-bomber showed what was possible. With almost identical engine power, the Whirlwind carried the same bombload as the Blenheim, packed a mighty punch with its concentrated pack of four 20 mm cannon and was capable of 303 mph at sea level (faster than the 283-mph Bf 109 or the 282-mph Spitfire I!) or 351 mph at 15,000 ft (faster then the 338-mph managed by the Bf 109 or the Spitfire's 342 mph). Its range, at 720 miles, was admittedly 400 miles less than that of the Blenheim, but this was almost academic, since the Blenheim was too vulnerable to actually make use of its 1125-mile range. A Whirlwind with twin Merlins might have been even more impressive.

Had squadrons of Whirlwinds attacked some of the targets against which whole units of Blenheims were lost, the targets would have been hit more heavily, the RAF would have suffered minimal casualties, and defending fighters might have taken a 'pasting' too. Alternatively, over shorter ranges, aircraft like the Hurricane could operate with more chance of survival as fighter-bombers, packing much the same punch, but with superior speed and agility.

But at the end of the day, the RAF had to go to war with what it had, and not with what it could (or should) have had. And by comparison with the RAF's other bombers, the Blenheim was very good indeed. Within weeks of the beginning of the 'Phoney War', heavy losses had virtually

The cockpit of a Blenheim I, with the observer's seat folded down beside the pilot's. No bombsight is fitted, and the aircraft has no ring-and-bead gunsight, indicating that the photo was taken at the factory, before completion. The aircraft has a full blind-flying panel, and has an old-fashioned 'flat' compass. The extent of the aircraft's nose glazing is readily apparent (*Aerospace Publishing/Wings of Fame*)

confined Bomber Command to night operations, and it was felt that only the Blenheim was fast and survivable enough to operate by day. Although many Blenheims were lost, the aircraft did manage to conduct meaningful operations, 'hitting the enemy where it hurt' and foreshadowing later, better known, daylight raids by aircraft like the Douglas Boston and the de Havilland Mosquito. And as a result, the Blenheim was popular.

If you didn't know any better, the Blenheim seemed an attractive and impressive machine, especially by comparison with the lumbering Whitley. Chris Patterson described his first sight of a Blenheim as 'the most charismatic and glamourous thing I'd ever seen', whereas of the Whitley he said, 'I'd never seen such a dreadful, boring looking, thing, nose down, going at what looked like 50 miles an hour. I found that flying it was exactly what I'd dreaded; it was slow and cumbersome, heavy and unresponsive.'

Finally posted to the Blenheim, Patterson was overjoyed, 'I opened the throttles and felt this surge of power, and the taking off, the lightness – I knew it was for me, that this was absolutely my plane'. Faced with the stark choice between the Whitley, the Wellington, the Battle and the Blenheim, it is no wonder that competition to be posted to a Blenheim squadron was fierce.

And it must be remembered that in the mid-1930s, when the Blenheim took shape, the RAF's senior commanders were perhaps a little hidebound, and were perhaps not completely open to revolutionary new ideas. This was an era in which many senior officers still believed in the open cockpit for optimum all-round visibility, and was an era in which the biplane fighter still had its adherents. A new RAF light bomber was expected to have a crew of three, with a dedicated gunner handling a single 0.303-in rifle calibre machine gun. Mounting this in a power-operated turret was quite revolutionary enough, without doing away with the weapon altogether! A Hind replacement would obviously be a relatively lightweight machine, capable of operating from delightful little grass airfields like Andover, Bicester or Netheravon.

Many were unconvinced as to the merits of any Hind replacement being a monoplane, let alone one with retractable undercarriage, and other such 'fripperies'. Moreover, the 1933 Air Exercises seemed to have shown that unescorted formations of light bombers, flying in neat boxes and covering one another with their rifle-calibre machine guns, would

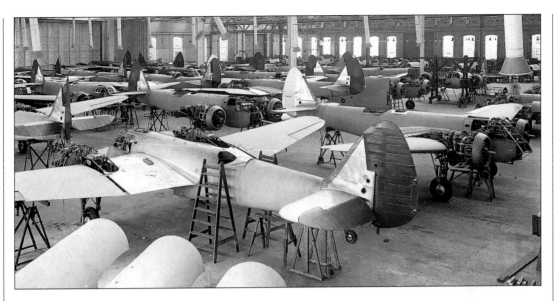

'always get through'. The Umpires (often flying aboard the bombers) believed the two, three, four or five men in a bomber crew when they claimed that they would have shot down all the attacking fighters, but paid little heed to the claims of the individual pilots of the single-seat fighters. Moreover, the bombers flew over their targets, and the assumption was made that any bombs dropped would have hit, and that any hit would have caused unimaginable destruction. Many didn't regard speed as being a particularly important consideration – what they felt was important was bombload and defensive firepower (the latter measured in numbers of inadequate 0.303-in machine guns!).

But there were those with a clear view of what was required. Such people included the crusading newspaper magnate, Viscount Rothermere, proprietor of the *Daily Mail*, who provided the initial impetus for what became the Blenheim. The Bristol Aeroplane Company originally began working on a light commercial aircraft, capable of a 250 mph cruising speed, in July 1933. This, the Bristol 135, was to carry two pilots and six passengers, and was to be powered by two 500 hp Bristol Aquila engines. Rothermere was a great patriot and a strong advocate of British aviation, and ordered a 240-mph, Mercury-engined Bristol 135 for his own use, stung into action by claims for the American Douglas DC-1, and in the knowledge that Lord Beaverbrook, his great rival, wanted to order one of the new Douglas transports.

Rothermere wanted the fastest commercial aeroplane in Europe, and he wanted it to be British. Bristol accepted his order after confirming that the Air Ministry (Bristol's main customer) would have no objection, worried that Rothermere would use the aircraft to point to the inadequacy of the RAF's existing fighters and bombers (which was his secondary motive in ordering the aeroplane). Far from objecting, the RAF decided to watch the aircraft's development with great interest. The Mercury-engined 135 became the Bristol 142, and a second, slightly enlarged, eight-seat prototype, powered by Aquilas, was built in parallel as the Bristol Type 143.

Foreign interest in a militarised version of the Type 143 was shown by Finland and a number of other countries. This, the Type 143F, was a

Blenheim Is in production at Filton. At least 40 aircraft are visible in the original photo, and most of those which have their control surfaces fitted have had the latter already doped in the finished camouflage colour. They would then be masked while the rest of the airframe was painted to prevent any build up of paint on the fabric-covered rudder, ailerons and elevators. Production of the Blenheim expanded extremely rapidly, allowing for a swift enlargement of Bomber Command, whose pre-war build-up was virtually centred on the Blenheim (*via Aeroplane*)

transport, but had provision for interchangeable nose sections, one with a fixed forward firing 20 mm Masden cannon and a free-mounted Lewis gun in the dorsal position. Negotiations for nine Type 143Fs began in February 1935, and on 12 April 1935, Rothermere's Type 142 made its maiden flight. Early flight tests showed great promise, and the Chief of the Air Staff wrote to Bristol, stating that the aircraft could be considered as a medium bomber 'if Bristol have a reasonable proposition to put forward for the supply of this type in reasonable numbers' and offering 'to test the aircraft made for Lord Rothermere at Martlesham free of charge in order to ascertain its performance and characteristics'.

This might seem to be lukewarm enthusiasm, but it must be remembered that the RAF's newly formed Bomber Command was concerned more with striking at the enemy's 'nerve centres' than with tactical bombing, and with carrying on the traditions of the Great War's Independent Force, which had attacked German cities with the aim of attacking 'root industries' and the morale of the enemy nation. Bomber Command was not really interested in medium bombers (let alone lightweights like the Blenheim). What it wanted was long-range 'heavies' with which to attack the Ruhr and German industrial centres further east.

When operational plans were drawn up in 1937, these included plans for attacks on aircraft manufacturing plants, for closing the Kiel canal, attacks against 'manufacturing resources' in the Ruhr, Rhineland and Saar, and elsewhere, and even for attacks on 'headquarters and administrative offices in Berlin and elsewhere'. Electrical power stations, coking plants, and even the Ruhr Dams were thought to be the vital targets. True strategic bombing was felt to be more useful and more realistic than tactical

Blenheim's lined up outside Filton's flight sheds, awaiting their pre-delivery test flights. The large size of the Blenheim can be gauged by the size of the groundcrew, and the trilby-wearing test pilot in the cavernous cockpit of the nearest aircraft (*via Aeroplane*)

One of the first Blenheim IVs pictured at Filton in front of a Blenheim I. The way in which the port side of the top of the nose is 'scalloped away' can be clearly see from this angle. This improved the pilot's view over the nose, while giving the observer more space in which to work (*via Aeroplane*)

bombing of hard-to-find, hard-to-hit and easy-to-repair targets.

The Blenheim's relevance to such plans seemed marginal (except in producing aircrew for new aircraft which were not yet available, and except overseas, where the Colonial policing type role was still best served by light bomber types). Many still believe that the Blenheim's greatest service was in equipping the squadrons which would later operate the Lancaster, Halifax and Mosquito, training the aircrew who would later fly these aircraft, and in providing experience and work for the factories which would later build the RAF's later bombers. When it emerged, the only real role perceived for the Blenheim was as a Hardy/Audax/Hart replacement for close support and army co-operation duties, and perhaps for colonial policing.

After naming the aircraft *Britain First*, Rothermere loaned the £18,500 aircraft to the Air Ministry

Bristol Aeroplane Company test pilot Bill Pegg is seen flying Blenheim IV L4842 near Filton on 29 May 1939. This aircraft subsequently went to No 53 Sqn, who kept it until it was lost in action on 17 May 1940. Before the outbreak of war, the Blenheim still seemed like an excellent aircraft, being both fast and agile (*via Aerospace Publishing/Wings of Fame*)

before it could take up its civil registration (G-ADCZ). It was sent to Martlesham Heath as R-12, where the RAF first requested an extended loan, and then asked to purchase the aircraft. Rothermere presented it to the nation free of charge, and it soon became K7557. Tests at Martlesham Heath demonstrated a top speed of 307 mph, or 285 mph with a full load. This was impressive stuff in 1935, and faster than the prototype Gladiator prototype, although the definitive Blenheim bomber was still a year from flight and almost two years from service, and no production Blenheim ever reached such speeds!

Bristol proposed a military version of the aircraft as the Bristol 142M, with either Aquila or Mercury engines. The bomber featured a modified nose containing a bomb-aimers position, a raised wing to allow an internal bomb bay and, most controversially, a powered dorsal gun turret. This last measure reduced the top speed to an estimated 262 mph with Aquila engines, or 278 mph with Mercuries. Specification 28/35 was issued to cover the Mercury-engined aircraft, 150 of which were ordered 'off the drawing board' in September 1935. This killed off Finland's Bristol 143, since RAF needs took precedence, though the country did purchase (and then licence-build) Blenheims. The sole Bristol143 (R-14) became an engine testbed for the Aquila, and Rothermere's order for a replacement 142 had to be turned away, such was the sudden priority given to producing Type 142Ms for the RAF!

Hitler had come to power in Germany in 1933 (but had led the largest party in the 1932 Reichstag, and had come second to Hindenburg in

Blenheim Is of No 114 Sqn – the RAF's first Blenheim unit – lined up at the 18th RAF Display at Hendon, during June 1937. The aircrafts' red '114' codes can just be discerned on the rear fuselage, the numerals being thinly outlined in black, with white individual code letters on the nose. The use of the squadron number as an identity code was routine before World War 2 (*via Phil Jarrett*)

Presidential elections in that year) and had given himself dictatorial powers. After 1935 there was little excuse not to see in which direction Germany was travelling, since the country denounced the Versailles Peace Treaty, introduced military conscription, openly established an air force and implicitly rejected disarmament. But in Britain, memories of the slaughter of World War 1 meant that rearmament was undertaken half-heartedly and without enthusiasm. But the Blenheim was at the heart of what rearmament there was as it gathered pace.

When it first flew on 25 June 1936, the first Blenheim I (K7033) had only a dummy turret fitted, and underwent service trials at Martlesham Heath in this configuration, being cleared for full-scale production in December 1935. This was not really a prototype, since the Type 142 had (it was felt) already amply demonstrated the type's potential. By the time the first Blenheim Mk Is were delivered to No 114 Sqn in March 1936 they were fitted with type B.I Mk I turrets, had controllable cooling gills around the engine cowlings and had their tailwheels locked down. Otherwise they differed little from K7033.

And if the production Blenheim was significantly slower than the prototype Type 142 had been two years earlier, no-one seemed to mind, and no-one seemed to notice that instead of Gladiators and the like, modern fighters were now exemplified by the Spitfire and the Bf 109. The new bomber would have two engines, dash it all, and promised to carry twice the bombload of the Hind more than twice as far, and to the Air Staff this seemed quite revolutionary enough.

The Air Staff's idea of a modern bomber was typified by the Vickers Wellesley, a draggy, fabric-covered, monoplane which first flew in 1935, and entered service in 1937. Ironically, the Blenheim I's cruising speed was only 12 mph higher than that of the Wellesley and its defensive armament was identical. The Wellesley actually carried double the bombload and with its geodetic structure was probably more resilient to ground fire. Such was the 'advance' offered by the RAF's new bomber. Thus while the Blenheim seemed terribly futuristic, marking a revolutionary advance over the aircraft it was replacing, the 'cutting edge' of bomber design was actually already further ahead of the Blenheim than the Blenheim was ahead of the biplane bombers!

It might all have been less embarrassing had the Blenheim had adequate armour protection and defensive armament, but with a single forward-firing Browning and a single turret-mounted Vickers machine gun, the aircraft was quite incapable of defending itself. And no-one had foreseen the host of other weaknesses which would afflict all of the RAF's first generation monoplane bombers. So, like the Wellington, Whitley, Hampden et al, the Blenheim entered service with obsolete radios, outdated and largely useless navigation equipment, an almost complete lack of armour protection, an unreliable oxygen system and no efficient heaters.

And as if all that were not enough, the Blenheim carried outdated, inefficient and largely ineffective bombs, and had a bombsight which was at best primitive. The result was that many Blenheims were lost because they could not fight off attacks, or because they could sustain so little punishment when attacked, and even more were lost because their crews got lost, or suffered from anoxia or lost concentration in the bitter cold. Crew comfort might seem a trifling matter, but cold and fatigue killed all too often. Blenheim crews had to fly encumbered in multiple layers of heavy flying clothing, and with crystal-based 'Everhot' bags strategically stuffed down jackets and trousers. All of this could cause problems not anticipated by Bristol's designers, as Sgt Les Spong of No 139 Sqn related:

'The cold encouraged the working of the urinary system . . . There was a relief tube . . . but I found that with all the layers of clothing . . . and the shrinking effect of the cold on one's tender part, it was impossible, using one hand only, both to make the part in question see the light of day and to hold the relief tube in position.'

Poor Spong had to hold the relief tube and get his long-suffering observer to 'perform the more delicate part of the operation!'

Most of the early Blenheim squadrons had previously flown variants of Hawker's Hart and Hind, compared with which the Blenheim must have seemed the very epitome of modernity. Here a No 62 Sqn machine cruises at close to maximum speed, its dorsal turret in the retracted position in an effort to decrease drag. Like the biplanes it replaced, the Blenheim originally carried its serial number both on the rear fuselage and on the rudder

Pilot (wearing a seat-pack parachute) and observer disembark from their No 90 Sqn Blenheim I at Bicester during the 1938 Annual Air Exercises. The gunner (inevitably a part-time member of the crew during this period) has already divested himself of Mae West and parachute harness, and is sitting in the rear hatch (*via Aeroplane*)

In addition, aircraft could soon become heavily iced, and could then stall and spin. Sometimes crews returned having recovered from such spins, but many others must have bulked out the 'missing' and 'failed to return' statistics. And all too often when they reached their targets, the Blenheims could do little damage. British GP (General Purpose) bombs had a pathetically low charge-to-weight ratio of only 27 per cent (German bombs were closer to 50 per cent) and were in any case very small. Moreover, an unacceptably small proportion of bombs dropped actually exploded. The Blenheim's inadequate warload would typically consist of 20 40-lb anti-personnel bombs and two 250-lb GP bombs. These latter had to be dropped from 1000 ft or higher, since if they did explode, the blast effect could damage the bomber if it was flying any lower! 1000 ft was pretty well perfect for anti-aircraft gunners, and gave the enemy sufficient warning of the Blenheim's approach. It was not low enough to make life difficult for Messerschmitt pilots either.

But as the need to expand became increasingly apparent, such weaknesses were ignored, and in July 1936 the RAF ordered an astonishing 434 more Blenheim Mk Is, later raising this total by another 134. Subsequently contracts were placed with Avro for 250 Blenheims to be built at Chadderton, and with car-makers Rootes for 600 aircraft to be built at Speke. Some 1568 Blenheim Is were eventually ordered for the RAF, some of which were diverted to Finland, Greece and Yugoslavia. Perhaps the need was felt to be for an interim, semi-modern, bomber for peacetime training, with the plan being that the Blenheims would be ditched in favour of something better (perhaps the new 'heavies') before war could break out. Tragically, this was not to be the case.

Even if the Air Ministry had failed to take notice of any intelligence reporting the performance of German fighters (difficult, when one of them had broken the world speed record, albeit in heavily modified form), then surely it might have safely assumed that the best German fighters were no slower than the Hurricane, even if they fell for the risible propaganda about the absolute superiority of the Spitfire. But apparently they did not, and Blenheim I production continued apace. Eventually some 694 Mk Is were built at Filton, including 18 for Finland, two for Yugoslavia, and 30 for Turkey, whilst Bristol's own line was augmented by Avro at Chadderton (250) and Rootes Securities at Speke (380). Licence production in Finland totalled 45 aircraft, and 16 more were completed in Yugoslavia – a further 24 were sabotaged on the line to prevent them falling into German hands when Yugoslavia was invaded.

Finland later received 24 ex-RAF aircraft, while Yugoslavia received 20 more Avro-built ex-RAF Blenheims. Six ex-RAF Blenheim Is also went to Greece, nine to Turkey and 39 (all but 19 being Rootes-built, the remainder being from Avro production) to Romania (including 13 Avro-built Blenheim Is sent as part of a failed diplomatic effort to prevent the coun-

Generations of aircrew have undertaken the faintly ridiculous charade of practising their ditching drill on dry land. Here the crew of a No 90 Sqn Blenheim I try to take it all seriously, as the gunner throws the dinghy pack onto Bicester's grass. The aircraft bears obvious traces of having had its A1 roundel toned down and converted to a B-type roundel. The steps for the aircrew are clearly visible above and below the wing trailing edge (*via Aeroplane*)

try from joining the Axis). One RAF Blenheim Mk I was evaluated by the SAAF, but was returned in August 1940 with a new serial number!

The massive production of the Blenheim I allowed a huge and rapid expansion of Bomber Command, and by the time of the Munich Crisis in September 1938, Blenheim Is equipped seven squadrons of No 1 Group, six more in No 2 Group and four in No 5 Group. No 1 Group subsequently divested itself of its Blenheim units before taking its Battles to France as the Advanced Air Striking Force, four going to No 2 Group and two to No 6 (Training Group), along with two No 2 Group units. One more No 1 Group squadron went to the Far East Air Force, along with one of the No 2 Group units. No 5 Group converted to the Hampden, although one squadron retained Blenheims and transferred to No 2 Group.

A new variant of the Blenheim was ordered as an interim general reconnaissance type for Coastal Command to meet the urgent need for an Anson replacement. This took over the Type 149 designation once applied to a Blenheim-based forerunner of the Beaufort, but was rather less radically modified, and was initially named Bolingbroke I. The new variant had an extended forward fuselage to allow for a prone bomb-aiming position, and had modified outer wings to allow increased internal fuel tankage.

The original long nose (fitted to K7072 – the sole Mk III prototype) was similar to that of the Blenheim Mk I in outline, but this placed the windscreen too far from the pilot's eyes. The nose was redesigned, with a normal close screen, and with the roof of the navigator/bomb-aimers compartment in front of him lowered and scalloped to avoid obstructing his view. The new long nose was fitted to K7072, which thereupon became the prototype

Bicester-based No 90 Sqn Blenheims are seen airborne in strength on 11 November 1938. The aircraft all wear post-Munich codes, with toned down roundels, and all have their turrets in the extended position. Overwing roundels show obvious traces of having had their yellow rings over-painted
(*via Aeroplane*)

This snow-bound Blenheim I is seen at Waddington in the late winter of 1938 wearing the post-Munich code letters of No 44 Sqn. The original roundel and underwing serials are just still visible, despite having been over-painted. The Blenheim formed the backbone of Bomber Command during the expansion, and in the post-Munich period
(*via Andrew Thomas*)

No 53 Sqn lined up its brand-new Blenheim IVs for the press after forming at Odiham in January 1939 as the first Blenheim IV unit. One aircraft in the line-up already has A-type roundels, whilst three do not have their Munich-era codes – the pre-war 'TE' code soon gave way to the code prefix 'PZ'. The nearest aircraft in the line-up, L4841, failed to return from a sortie on 19 May 1940 (*via Phil Jarrett*)

Blenheim IV. The name Bolingbroke was then abandoned (though it was revived for the aircraft built under license by Fairchild Aircraft in Canada) in favour of the Blenheim Mk IV designation.

Although designed to meet a Coastal Command requirement, the Blenheim IV marked enough of an improvement over the Mk I to be attractive to all Blenheim operators, and the new version simply replaced the earlier variant on the production line, for delivery to Bomber Command, overseas units, and Army Co-operation Command as well as Coastal Command. There were even fighter conversions of the Mk IV. Blenheim IVs rapidly replaced Blenheim Is in home-based RAF squadrons, making the Mk Is available for training, fighter conversion or overseas use.

Bristol built 313 Blenheim IVs at Filton, including a batch of 12 for Greece. Two Bisley (Blenheim V) prototypes were also built by Bristol – this final army support variant is described in more detail elsewhere in this book. As with the Mk I, the Blenheim IV was also built by Avro, who produced 750 aircraft, and by Rootes, whose Speke and Blythe Bridge production totalled 2100 before the company switched to the Blenheim V, 942 of which were constructed at Blythe Bridge.

A further 656 Blenheim IVs were built by Fairchild in Canada as Bolingbrokes, most being US Pratt & Whitney Twin Wasp Junior-engined Mk IVs – the first 18 were Mercury VII-engined Mk Is, however, and the last 57 were powered by Mercury XXs.

Ten Mk IVs were completed in Finland, but five incomplete aircraft were scrapped when the Finns finally made peace in 1944. At least 17 ex-RAF Blenheims (mainly Mk Vs) were delivered to Turkey, augmenting that nation's Mk Is. Portugal also became a Blenheim operator, mainly by flying interned Mk IVs which had force-landed, but also by acquiring ex-RAF Blenheim Vs. The very large production total did not reflect complete blindness to the Blenheim's faults, however. The aircraft was easy to construct, and no alternative aircraft type was ready for large scale production at that time. One has to ask whether a massive increase in the number of Whitleys, Wellingtons or Manchesters would have improved the RAF's fortunes during the first year of the war?

When war broke out, the RAF had some 168 Blenheim IVs on charge, including three with a Fighter Command nightfighter squadron (No 25), and 33 with two Army Co-operation units of Bomber Command (Nos 53 and 59). The rest equipped seven frontline bomber squadrons of Bomber Command, Nos 82, 90, 101, 107, 110, 114 and 139, while No

No 53 Sqn belonged to Army Co-operation Command, and operated in the strategic reconnaissance role. The unit went to France with the Air Component of the BEF, and played a major part in the Battle of France. This aircraft (L4843) was shot down by groundfire over Belgium on 16 May 1940 (*via Bruce Robertson*)

21 Sqn was on the verge of re-equipping with the new variant. These squadrons all served with No 2 Group, at Wyton, Wattisham, Watton and West Raynham, excepting No 90, which was a No 6 Group pool squadron based at Upwood. No 6 Group also had four squadrons of Blenheim Mk Is, two of which immediately transferred to the Air Component in France. Sixteen of Bomber Command's 33 squadrons flew Blenheims as war broke out.

But the large number of squadrons gives an entirely false impression of the RAF's real fighting strength and preparedness for war on 3 September 1939. Long-serving squadrons were split to form new units, with a flight of one squadron (perhaps itself newly formed) becoming the cadre for a new unit. The experienced, long-serving, regular pilots and aircrew were widely scattered, forming the backbones of the new expansion-period squadrons, which were mostly manned by very inexperienced, barely trained, and very young Volunteer Reservists.

The bulk of any given squadron would be formed from very junior pilot officers and sergeants, the majority of whom were fresh from flying training. Such was the shortage of more experienced men that one flight commander would usually be a flight lieutenant, but the other would be a flying officer, or even the most senior of the pilot officers (who, by definition, would have less than two years service behind him). In some squadrons even the senior flight commander would only be a flying officer. Often the most experienced pilots on a unit were the sergeants, recruited from the other ranks and destined to return to their old trades after seven years flying, having effectively gained accelerated promotion to sergeant through having been pilots. Many observers felt that the RAF's frontline squadrons were actually less well prepared for war than were the Auxiliaries, which at least still had their long-term commanders and flight commanders in place.

Great efforts were made to improve the Blenheim during the early months of the war, with the adoption of a new colour scheme which used lower-drag pale duck egg blue Camotint paint, and with the adoption of a twin Vickers K-gun mounting in the mid-upper turret. There was also a slow adoption of a rearward firing gun in a transparent blister below the nose, though this remained rare until after the Battle of France.

'PHONEY WAR' AND *BLITZKRIEG*

The Blenheim actually performed the first RAF sortie of the war when Flg Off McPherson flew Blenheim IV N6215 of No 139 Sqn from Wyton to Wilhelmshaven, where his observer, Cdr Thompson, RN, photographed enemy naval units as they left the harbour, from 22,000 ft. The radio froze and was unusable, but fortunately gunner Cpl Arrowsmith was given nothing to do by the Luftwaffe, and even the enemy flak units were silent. The Blenheim successfully returned to base at 1650. The crew had been on standby during the two previous days, but had not been given the order to launch. On 3 September, just such an order was received one minute after Prime Minister Neville Chamberlain's announcement that the country was at war, and as a result, the aircraft arrived home too late for its photos and crew reports to be used to allow the launch of a bombing raid that afternoon. Accordingly, a second recce mission was mounted the following day, again flown by McPherson. This time cloud cover forced him to overfly his objectives at 250 ft, returning to base after a gruelling four hours. These historic first sorties of the war (and subsequent missions) won McPherson the DFC.

It is perhaps remarkable that the pilot and his crew even found their target, such was the poor state of Bomber Command's navigation. Aircraft carried an astro-sextant (of course!), but relied on dead reckoning and only the most primitive direction finding. Navigation had been the responsibility of the pilot until 1937, when second pilots, gunners and observers were given short navigation courses. It was not until 1938 that the need for a dedicated, fully-trained, navigator was recognised, despite the fact that operations against the likely enemy would entail long missions, much of which would be flown over the featureless North Sea.

Thus even as the war began, many Blenheim sorties were made by a

Pictured late in 1939, this No 82 Sqn Blenheim IV was one of those used in No 2 Group's initial campaign of armed reconnaissance sorties over the North Sea and along the German coast. These early sorties predated the introduction of higher conspicuity roundels in October 1939 (ordered in a vain attempt to prevent losses to friendly fighters) and the adoption of light undersides. Only the use of 'UX' (rather than pre-war Munich 'OZ') codes reveals that this is a wartime photograph (*via Andrew Thomas*)

fully-trained pilot, a corporal who had undergone a brief training course acting as observer and with a part-time gunner, probably one of the squadron's fitters or riggers, who had undergone a brief course, and who got flying pay when he flew, and wore the winged bullet badge on his sleeve. He would be a far cry from the air gunners of later years, who were sergeants like the sergeant pilots, and for whom air gunnery was their sole duty, and for which they had been rigorously trained. A few lucky crews had full-time, fully-trained navigators, since these had started to be trained from June 1938.

For the first months of the war, Bomber Command was restricted to attacks which were of peripheral importance, at best. The supposed German bombing raid on Guernica shaped public and political hostility against attacks on cities, as did the fear of German retaliation. Unrestricted air warfare was not deemed to be in Britain's interests, and the policy was to let Germany set the

The presence of belted ammunition reveals that this Blenheim IV has the later twin-Browning dorsal turret. The original Vickers Gas-Operated (VGO) Type K machine gun used pre-loaded circular drums of ammunition, like a World War 1 Lewis gun. This aircraft also appears to have a free-mounted 0.303-in Browning in the nose glazing

agenda ('they have the initiative, owing to their superior strength'). The Air Ministry elaborated three general principles – that the intentional bombing of civilians was 'illegal', that identification and distinguishing of a target was a prerequisite to an attack and that 'Bombardment must be carried out in such a way that there is a reasonable expectation that damage will be confined to the objective . . . civilian populations in the neighbourhood will not be bombarded through negligence'.

Attacks on targets which might entail indirect attacks on enemy civilians were off-limits until Winston Churchill intervened in the summer of 1940. Attacks on the Ruhr were ruled out by these constraints, and in recognition of the fact that Bomber Command was still too weak to mount anything but a sporadic and somewhat desultory campaign, which would be largely ineffective. Bomber Command's activities were thus restricted to the night delivery of propaganda leaflets, and to operations against the German coast and coastal area, primarily aimed at dispersing fighter defences between Germany and the Czech and French fronts!

On 4 September, using the recce photos taken by McPherson and his crew on the first day of the war and further photos taken that morning, 15 Blenheims were among the fleet of 29 aircraft tasked to attack German warships at Wilhelmshaven – 'The greatest care is to be taken not to injure the civilian population', read the Operation Order. 'The intention is to

Groundcrew load 125-lb bombs into the bomb bay of a No 2 Group Blenheim – the device to the left is an armour-piercing weapon, with its fuse at the tail. The effectiveness of the Blenheim was severely constrained by the poor reliability of the bombs it carried, which had a low ratio of explosive to overall weight, and which were notoriously unreliable (*Aerospace Publishing/Wings of Fame*)

Passed by the censor for first publication on Friday 22 August 1940, this photo showed, according to its wartime caption, 'A formidable egg in its nest in one of the bombers taking part in the sweep off the Dutch coast, seen before setting off. This beautiful bomb wrecked (sic) much havoc when this plane reached its target' – if indeed it reached the target, and if it exploded (*Aerospace Publishing/Wings of Fame*)

destroy the German Fleet. There is no alternative target'. No 139 Sqn's five aircraft turned back before reaching the target, deterred by very poor weather conditions. No 110 Sqn did better, led by Flt Lt (later Sqn Ldr) Ken Doran. The first wave attacked successfully, losing only one aircraft. Four Blenheims from No 107 Sqn (the second wave) were shot down as they attacked warships in Wilhelmshaven harbour, although one crashed into the forecastle of the destroyer *Emden*. The force did put two (and according to some sources at least three) 500-lb bombs into the *Admiral Scheer*, although none exploded. Five of the 14 Wellingtons accompanying the Blenheims also turned back, and two were downed by Bf 109s, whose shooting was described as 'wild', and which failed to press home their attacks, otherwise the cost might have been higher (see *Osprey Aircraft of the Aces 11 - Bf 109D/E Aces 1939-41* for further details). Thus ended Bomber Command's first offensive mission of the war.

Blenheim solo reconnaissance missions against the German navy and the northern German coast proved costly, with a loss rate approaching 20 per cent, and these were eventually abandoned in late November. Before the war it had been assumed that a ten per cent loss rate would effectively render a unit incapable of further participation in a particular battle, requiring rest, reinforcement and recuperation. Blenheim units suffered ten per cent casualties on a regular basis, yet remained in the line. Anti-shipping searches and strikes proved little more successful, most returning to base without finding the enemy. When enemy aircraft were encountered, things tended to go badly. On 10 January 1940, for example, four Bf 110s of 2./ZG 76 tangled with nine No 110 Sqn Blenheims, shooting down one and causing such damage to two more that they were written off after landing. Fortunately the German pilots' shooting was wild, and they returned to base after expending all of their ammunition.

There were exceptions to the Blenheim's run of poor form, of course, including one memorable mission on 11 March when a No 82 Sqn Blenheim (P4852) flown by Sqn Ldr Paddy Delap sank the U-31. This was a particularly fine achievement, since a U-boat is a small and mobile target, and as late as 1938, Bomber Command had been having difficulty getting a 15 per cent accuracy rate against static, undefended, targets from medium level and a three per cent accuracy rate in high level bombing. The U-31 was undergoing post-refit trials in the Jade Bight when it was attacked on the surface by Delap's Blenheim, which hit the vessel with two impact-fused bombs. The U-boat sank in 50 ft of water and with the loss of all 58 aboard (including 13 dockyard workers), but was later raised, repaired and recommissioned. The low-flying Blenheim was itself damaged by shrapnel and splinters, and by the bomb's shock-wave, but made it back to base, becoming the first aircraft to sink a U-boat unaided.

But if the Blenheim raids generally had little material affect on the enemy, they did serve at least one useful purpose. RAF Fighter Command had been assumed to be powerless against incoming mass raids by German bombers (on the basis of the infamous 1933 Air Exercises), and yet the success of German fighters against British bombers showed that the bomber was vulnerable. Fighter Command could thus face the future with greater confidence.

As war broke out, the RAF included eight frontline Blenheim IV bomber squadrons (Nos 21, 82, 90, 101, 107, 110, 114, and 139 Sqns), with two more Blenheim IV units operating in the army co-operation role. Two more bomber units (Nos 18 and 57 Sqns) operated the Blenheim I, but these were soon transferred from No 1 to No 6 (Training) Group. They later transferred to the Air Component of the BEF, and while with the latter force they rapidly re-equipped with Blenheim IVs. Prior to the German invasion the two squadrons operated principally in the strategic reconnaissance role. Attrition was high, and by late 1939 all replacement aircraft were Mk IVs.

A British Expeditionary Force (BEF) was sent to France on the outbreak of war, and this had its own Air Component, which eventually (by the time the Germans invaded) included four Blenheim squadrons – Nos 18, 53, 57 and 59 – as well as Lysander and Hurricane units. The BEF was also accompanied by an Advanced Air Striking Force (AASF) with ten squadrons of Battles, this reporting to AOC-in-C Bomber Command. In December 1939 two of the AASF Battle squadrons (Nos 15 and 40) rotated home to re-equip with Blenheims, and were immediately replaced by the Blenheim-equipped Nos 114 and 139 Sqns. The Air Component and the ten-squadron AASF (under direct Bomber Command control) were effectively merged when all RAF assets in France were brought under the command of British Air Forces in France.

Thus, while home-based Blenheim squadrons harassed the 'Hun' at sea, and in his ports and coastal regions, some six Blenheim squadrons were soon to taste action over France and Germany flying from French bases. These aircraft started taking over the brunt of reconnaissance missions, since the Battles proved terrifyingly vulnerable to both flak and fighters, and large numbers were also lost in accidents whilst attempting to fly ever lower in poor weather or at night.

Even before war broke out, it was apparent that the Blenheim would be hacked down in huge numbers if it tried to operate by day, in the face of fighter opposition. Logically, the aircraft should have been used by OTUs and perhaps at night, or in theatres or circumstances where enemy fighters would not be encountered. Using them by day had only one purpose, and that was political – proving that the RAF had the means of attacking enemy targets 'round the clock', and demonstrating a commitment to the defence of France, even if that prejudiced the later defence of Britain itself. But to do so at such heavy cost represented a triumph of bravado over reason, and marked a callous willingness to sacrifice valuable aircrew for nothing. Even Göring never asked the Luftwaffe for such pointless sacrifices. But while the rest of Bomber Command switched to night bombing, No 2 Group continued to mount its missions in the full glare of daylight, and continued losing Blenheims and their crews at an alarming rate.

On 6 April 1940, Basil Embry led his No 107 Sqn in an attack against the *Scharnhorst*, *Gneisenau* and a convoy heading north for the invasion of Norway – *Scharnhorst* was hit, but the bombs failed to penetrate the battleship's armour. The German invasion of Norway was hardly affected, and a new campaign opened for the Blenheim, one which was to be long, drawn out and extremely costly. Even after Norway fell, bombing attacks continued, aimed primarily at halting the export of Norwegian iron ore to Germany, most of which left Norway via Narvik. No 2 Group's Blenheim squadrons began rotational detachments to Lossiemouth for

A Blenheim gunner poses for the official camera, crouched behind his VGO machine gun. Even with a power-operated turret, a single 0.303-in gun was little better than a peashooter against a 20 mm cannon armed Bf 109 (*Aerospace Publishing/Wings of Fame*)

operations off Norway almost immediately.

It is uncertain as to whether night operations would have been much more preferable. Bomber Command had done little training at night before the war, and problems of navigation and bombing accuracy would have been intensified. Sir Lewis Hodges (later Marshal of the RAF) recalled that his introduction to night flying in the Blenheim came six months after his single solo night circuit in an Airspeed Oxford during OTU training at Upwood;

It was a white Christmas for No 139 (Jamaica) Sqn at Bétheniville in December 1939. One of the squadron's Blenheim IVs is seen here shrouded in camouflage netting, photographed from underneath another camouflaged Blenheim. RAF air and groundcrew were prohibited from taking their own photographs, but this image was issued officially as a postcard for AASF and BEF use (*via Bruce Robertson*)

'I was told to take a Blenheim up with my air gunner and take-off at dusk and then keep going round and round the circuit until it was dark. You are in a completely strange and alien environment, doing everything entirely by instruments. Your whole technique of approach and landing is entirely different with a flarepath.'

Four aircraft were lost in a single night on Hodges' OTU course, and six of the 21 crews on the course were killed before they could reach a unit.

And while the Ministry of Information published pretty pictures of Blenheims setting out for another sortie against the 'Hun', complete with optimistic captions, few were not aware of the dreadful truth, and aircrew in other commands watched as the Blenheim squadrons of No 2 Group and Army Co-operation Command were decimated.

'Bristol should have delivered them before final assembly, and we could have piled up the wings and fuselages on the beaches as anti-invasion defences! They were worse than nothing at all, because they took valuable aluminium, engines, variable pitch propellors (VP props, for christ's sake, when Spits and Hurris still had fixed wooden props!) and production capacity and they achieved nothing at all. Nothing. Except that they killed the cream of the pre-war air force, guys who would have made a real difference had they been strapped into Spits, Hurricanes, or a modern bomber!" observed one former Hurricane pilot.

But doubts about the planned use of the Blenheim went to the very top. Portal himself, C-in-C of Bomber Command, was resolutely opposed to continued daylight raids by Blenheims, especially long range attacks against targets which were heavily defended by fighters. A 24 April minute to the Ministry outlined his objections in the strongest terms;

'It seems to me the height of folly to throw away experienced crews on the bombing of aerodromes which, I think you will agree, shows the least result for loss of equipment expended on it'.

Still unhappy about the way his Blenheims were being wasted, on 8 May 1940 – two days before the *Blitzkrieg* in the West began – he wrote;

'I am convinced that the proposed use of these units is fundamentally unsound, and that if it is persisted in, it is likely to have disastrous consequences on the future of the war in the air. It can scarcely be disputed that at the enemy's chosen moment for advance, the area will be literally swarming with enemy fighters, and we shall be lucky if we see again as many as half the aircraft we send out each time. Really accurate bombing

The wartime censor has deleted details of cable cutters on the leading edges of this No 107 Sqn Blenheim IV. The unit's Blenheims had a busy war, taking part in the first attack on Wilhelmshaven, and then in the Battle of France. The squadron transferred to Coastal Command for two months in March 1941, before enjoying a brief respite prior to being posted to Malta (*via Aeroplane*)

No 53 Sqn's L4860 was shot down near Cambrai on 16 May 1940, and then formed a macabre backdrop for this bizarre photograph of four young French girls. Within days their country would fall, and they would spend the next four years under German occupation. A shattered No 53 Sqn returned to the UK from France on 19 May 1940, but continued to fly strategic reconnaissance missions until it transferred to Coastal Command (*via Andrew Thomas*)

under the conditions I visualise is not to be expected, and I feel justified in expressing serious doubts whether the attacks of 50 Blenheims based on information necessarily some hours out of date are likely to make as much difference to the ultimate course of the war as to justify the losses I expect.'

But apart from furiously minuting the Air Ministry, there was little Portal could do other than to obey his orders and watch impotently as those orders directed his Blenheims off on the most lunatic missions. To rub salt into the wound, he then had to sign his name to the routine congratulatory telegrams and signals intended to pep up the morale of the very units he was watching bleed dry.

But from 10 May 1940 things changed. Rested, re-armed and re-equipped after its invasion of Poland, the German Wehrmacht began a *Blitzkrieg* in the West, smashing through the Netherlands and Belgium and on into France. Instead of attacking over the Franco-German frontier (where it would have had to smash through the 'impregnable' Maginot Line), the Wehrmacht instead hooked down through neutral Belgium and the Netherlands. Unfortunately Belgium's neutral status led to delays in the British and French forces advancing to their pre-selected defensive line, which lay between the Meuse, Namur, the Dyle and Antwerp. Such was the confusion that on 10 May allied bombers were not given the necessary authority to overfly Belgian territory – a stricture not imposed on the Luftwaffe's *Kampfgruppen*!

One positive outcome of the *Blitzkrieg* was that the indecisive Chamberlain stepped down and Winston Churchill became Prime Minister. The gloves were off at last. Although the scramble to halt the German drive through France ensured that RAF aircraft would still be used wastefully against tactical targets, there was a new enthusiasm for attacks on targets in Germany itself. The bombing of Rotterdam by the Luftwaffe on 14 May convinced Churchill that the informal ban on bombing civilians was over, though indiscriminate attacks were still discouraged.

As the Wehrmacht romped

A rare and very illegal personal snapshot of a No 110 Sqn Blenheim in one of Wattisham's hangars, undergoing inspection. The airman who took the photo remarked that 'aircraft rarely survived long enough to be inspected', and N6214 proved to be no exception, being shot down off Norway on 25 April 1940, before its next scheduled inspection (*via Bruce Robertson*)

No 21 Sqn Blenheim IV N6166 is seen under repair at Bodney in June 1940, having limped back to base after receiving major battle damage. When Blenheim aircrew examined shot down He 111s they were shocked by the flimsy and light-weight construction of the German bomber. The Blenheim was well made and robust, and was able to survive a significant amount of battle damage. Unfortunately, the crew lacked armour protection, and were thus the 'weak link' in the chain. Had the Blenheim been less heavily built, it might have been fast enough to avoid some enemy fire (*via Peter H T Green*)

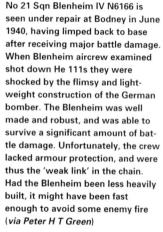

through northern France, the RAF was thrown into the fray to try to stop the advancing armour, and especially to destroy bridges across the rivers and canals so as to make them natural defensive barriers against the German advance. Most of these had already been planned for destruction, but the speed of the German advance (and bold use of paratroops) ensured that most bridges were captured before they could be blown. The only hope for the allies was to bomb them. Unfortunately, the German tank General Guderian brought up his AAA immediately, and the bridges were soon very heavily defended.

The obsolete Fairey Battles fell in droves, achieving little, and the Blenheims were unable to do much better, although the AASF and Air Component units were augmented by aircraft from No 2 Group, flying from their British bases. On 10 May, nine No XV Sqn Blenheims attacked Waalhaven aerodrome with some success (following a raid by six Blenheim fighters of No 600 Sqn, during which a Ju 52 was shot down), and 12 aircraft from No 40 Sqn attacked Ypenburg, two of the aircraft falling to Bf 110s. Finally, 12 Blenheims from No 110 Sqn attacked Waalhaven and its environs at 1650, claiming the destruction of five more Ju 52s. One No 2 Group Blenheim was lost on a recce mission that day, along with four Air Component Blenheims and five Fighter Command Mk IFs.

But if Blenheim losses were heavy on that day, they had hurt the enemy

badly. One *gruppe* of Ju 52s alone lost 42 of its 55 Ju 52s on the beaches and airfields of Ypenburg and Ockenburg, whilst other units suffered similar losses. Many of these were destroyed by Blenheims.

On 11 May almost all of No 114 Sqn's Blenheims were destroyed on the ground at Conde Vraux by Do 17Zs of II./KG2. Four more Air Component Blenheims were lost, along with two aircraft from No 110 Sqn. The latter unit had, however, managed to severely damage the bridge that had been its target, plus claim a fighter destroyed. Two crewmen from a downed Blenheim survived to become patients in a French hospital, although one was later made a PoW. No 21 Sqn returned from a 12-aircraft raid on German armour with all of its aircraft, although eight were subsequently declared unserviceable upon landing, and a gunner had been killed. Remarkably, no further attacks were made, such was the need to conserve aircraft and crews, even though German columns were speedily pushing south.

On 12 May, seven (of nine) No 139 Sqn Blenheims were shot down by Bf 109s of JG27 while attacking armoured columns and bridges (which had, in fact, already been destroyed) near Maastricht, having pressed on to the target despite the non-appearance of the scheduled fighter escort. Some 15 aircrew were killed. Among the casualties was Andrew McPherson, and N6215, the man and machine that had flown the RAF's first mission of the war. Although these were vital targets, the Blenheims were effectively flying a diversionary mission for the later attack by six Battles of No 12 Sqn assigned to attack bridges over the Albert Canal. The latter mission resulted in the RAF's first VCs of the war, but failed to destroy the targets. Later that day Basil Embry's No 107 Sqn lost four crews trying to bomb the Maas bridges, with No XV losing six more an hour later. Only two of the surviving aircraft could be repaired to fly again.

Two of eleven No 110 Sqn Blenheims were shot down on a similar mission, and one No 21 Sqn aircraft was downed during an evening mission. Bandon's No 82 Sqn were luckier, escaping losses during their evening sortie. However, the AASF and Air Component had seven more Blenheims destroyed that day. These losses prompted the Chief of the Air Staff to express his concern at 'these heavy losses incurred by the medium bombers. I must impress on you that we cannot continue indefinitely at this rate of intensity. If we expend all our effort in the early stage of the battle we shall not be able to operate efficiently when the really critical phase comes'.

Despite an official stand-down, four more Blenheims were recorded as missing in action on 13 May. But the next day was to be the worst day yet for the RAF in France, and a black day for the allies more generally. Heinkel He 111s of KG 54 took out the heart of Rotterdam, prompting the Dutch surrender. Also on 14 May a maximum effort was made against German bridges across the Meuse at Sedan which saw 44 of the 71 attacking aircraft

No 110 Sqn was one of the busiest of No 2 Group's Blenheim units, successively attacking airfields during the German invasion of Norway, then flying intensively in support of the BEF during the Battles for the Low Countries and France, before raiding invasion barges, flying anti-shipping missions off the enemy coast and detaching to Malta! These squadron aircraft are seen early in the war, probably at Wattisham. Both the aircraft shown were written off after hitting trees, demonstrating the emphasis placed on low level operations (*via Bruce Robertson*)

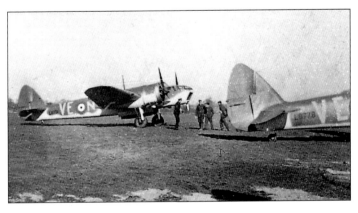

This No XV Sqn Blenheim was recorded as missing after attacking Le Cateau on 18 May 1940. The aircraft has the original transparent undernose fairing for a rearward-facing machine gun (which does not appear to be fitted), whilst its camouflage has a soft, unmasked, demarcation between upper and lower surfaces. RAF bombers were delivered without underwing roundels, which were introduced (along with red white and blue fuselage and overwing roundels) as a panic response to the loss of a Coastal Command Blenheim to friendly fighters (*via Bruce Robertson*)

lost, including five of eight Blenheims from Nos 139 and 114 Sqns. One Blenheim failed to return from a reconnaissance sortie, and another bomber was destroyed by fire by French civilians after it made a forced landing, the locals 'torching' the perfectly serviceable aircraft while the pilot and observer went searching for petrol! They had, however, claimed a Dornier Do 17 destroyed during their recce mission, although the aircraft's gunner had been killed in the engagement.

Six No 107 Sqn Blenheims attacked the advancing Germans without loss, but No 110 Sqn had less luck, despite a fighter escort, losing five of the twelve aircraft despatched, although they did also claim one Bf 109 shot down. No 21 Sqn lost one aircraft, but shot down at least two Bf 109s. This day marked the highest loss rate ever experienced by the RAF during World War 2, and while some bridges were destroyed, it was not in the end enough to seriously blunt the German advance. The Battles' losses were so heavy that they were effectively withdrawn from daylight operations. Interrogations of German prisoners did show that the RAF's sacrifices had not entirely been in vain, however, with evidence of severe disorientation and poor morale as a result of the regular air attacks. At one stage it seemed that the air attacks, and a French counter-attack on the ground, had actually halted the Wehrmacht's impetus, but unfortunately this proved to be only temporary.

Seven Blenheims fell the next day, too, two of them recce aircraft which were lost to RAF Hurricanes – the victims of misidentification. Two more were from No 40 Sqn, which mounted a combined raid with nine of its aircraft accompanied by three No XV Sqn Blenheims. The remaining casualties included one from four No 139 Sqn Blenheims which attacked Moutherme. No 82 Sqn bombed troop concentrations at Moutherme too, but without loss, the unit being led by the charismatic Earl of Bandon. 15 May also saw the beginning of a withdrawal of Air Component and AASF squadrons, many of which had to burn spares, stores and unserviceable aircraft as they fled. The Dutch surrendered the 24 hours later, but fighting continued to rage in France, although many date 15 May as the day on which France lost its war.

The RAF's reliance on delightful daisy-strewn grass airfields shaped a generation of combat aircraft, requiring a tailwheel undercarriage and relatively benign landing characteristics. In 1939, great emphasis was still placed on traditional flying skills, and a good landing was a three-pointer, preferably achieved without 'rumbling' (varying the throttle settings) on approach. This No 139 Sqn Blenheim tipped on its nose at Wyton during the first days of the war, probably after an over-enthusiastic application of the brakes (*via Bruce Robertson*)

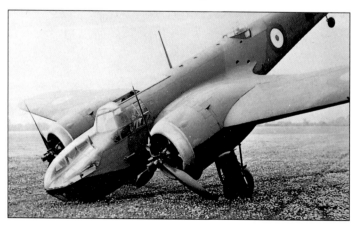

16 May saw two more attacks on Blenheims by RAF Hurricanes, one force-landing, wheels-up, the other being shot down. Two more reconnaissance Blenheims were downed during the day, but Flt Lt Simon Maude evened the score a little by downing two Luftwaffe Hs 126 observation aircraft using his fixed, forward-firing, gun. Finally, Nos 114 and 139 Sqns received orders to return to England. Everybody was becoming aware of the Blenheim's inadequacy, and on 16 May No 2 Group's SASO (Senior Air Staff Officer), Sir Hugh Pughe-Lloyd, commented in his diary that, 'We haven't got a bomber which can penetrate the enemy defences in daylight . . . we have no day bomber in our armoury'. If Pughe-Lloyd looked back at his diary he must have seen the irony, since two years later, on Malta, he was still being forced to rely on using Blenheims in daylight.

On 17 May 1940 No 82 Sqn suffered its greatest tragedy. The squadron's 12 aircraft, led by Sqn Ldr Paddy Delap, attacked German armour streaming through the Gembloux gap towards Mons, but were engaged by fierce AAA fire, which broke up the two boxes of six aircraft. Only one aircraft fell, but of the remaining 11, ten aircraft were quickly shot down by 15 attacking Bf 109s. Twenty aircrew were killed, three became PoWs, and others were injured. One of the survivors was the indomitable Delap, who found himself in a French hospital. Another survivor walked 60 miles to Paris, carrying his parachute all the way. The single remaining aircraft returned to Watton badly damaged, and was subsequently written off. Remarkably, the OC, Wg Cdr (later ACM) the Earl of Bandon, rebuilt the unit to fight again, as will be described, despite receiving orders to disband it. No 82 Sqn actually managed to mount sorties once again just three days later.

No 82 Sqn's misfortune that day overshadowed three more Blenheims lost on reconnaissance sorties, and with No 18 Sqn losing two of eight aircraft despatched. Losses continued to mount, No 2 Group seeing 44 aircraft destroyed in the nine days on which it operated up to 22 May, and with total Blenheim losses for the month exceeding 150 aircraft. At the end of the day, the Blenheim was not well suited to the close air support role, operating against the mobile targets offered by a rapidly advancing

Seen over the patchwork of tiny fields of northern France, this Blenheim has the new red, white and blue fuselage roundels in late 1939, but retains red and blue over-wing roundels. A directive to replace these had been issued in October 1939, although it theoretically applied only to reconnaissance aircraft. This No 139 Sqn Blenheim IV (N6216) was lost in action on 12 May 1940, when the fighting in France was at its fiercest. On that day, No 139 Sqn alone lost seven of the nine aircraft it despatched against targets near Maastricht. The aircraft still has its serial number repeated on both rudder and rear fuselage, following pre-war practise (*via Phil Jarrett*)

enemy army, but in truth, even a great CAS aircraft would have struggled in the face of overwhelming German air power.

On 18 May it was the turn of No XV Sqn, which lost three of six aircraft despatched, with two more being badly damaged. The survivors landed at Poix, where two were grounded, one of these being the aircraft of Leonard Trent (who later won the VC), who flew the damaged machine back to Mildenhall the next day. Seven aircraft sent out

by No 40 Sqn fared rather better, beating off attacks by Bf 109s without loss to themselves. Most squadrons had to abandon their damaged aircraft as the squadrons recoiled in the face of the lightning German advance, and virtually all the Blenheims returned to England on 19-20 May.

But that was far from the end for the Blenheim in the Battle of France. On 20 May Nos 21 and 107 Sqns each carried out two 'maximum efforts', heavily escorted by Hurricanes. The presence of a fighter escort was extremely welcome, and on each raid all the aircraft returned safely, although some were damaged. The same evening Nos 40, 82 and 110 Sqns each put up six Blenheims, all returning safely. The night raids achieved nothing, however, except to exhaust the already over-stretched 'Blenheim boys'. On 21 May Embry led 12 No 107 on a raid against German motor transports, escorted by Hurricanes, and again led all his aircraft home safely. No 40 Sqn managed a two-aircraft recce and a three-aircraft strike without loss, although No XV Sqn lost one aircraft (which made a successful belly landing and was burned by its crew) of six sent out. A No 18 Sqn aircraft despatched on a recce mission was lost around midday.

In the afternoon Nos 21, 82 and 107 Sqns mounted a combined mission, contributing nine, three and eleven aircraft respectively – all of which returned safely. Later that day a similar mission was flown by Nos 18, 82 and 110 Sqns (three, six and six aircraft respectively), with one bomber being shot up by six RAF fighters and belly landing to be burned by its furious crew. Six No 110 Sqn aircraft returned safely from the last Blenheim mission of the day.

Nos 57 and 59 Sqns each lost single recce aircraft on 21 May, but there were no offensive operations by No 2 Group, which had 73 aircraft available the following morning, and flew several attacks. One of six No 107 Sqn Blenheims had to ditch in the Channel on the way home, while No 110 Sqn lost one of six aircraft in the vicinity of its target. Another aircraft belly-landed upon returning to its base. Nos 15 and 40 Sqns each mounted operations which suffered no losses, while later attacks by Nos 21, 82, 107 and 110 Sqns resulted in only one loss.

On 23 May two solo recce Blenheims were downed, six aircraft each from Nos 40 and 107 Sqns suffering two losses despite their Hurricane escort, and in the afternoon only five of six No XV Sqn aircraft returned,

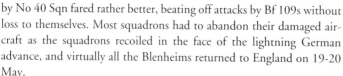

Tight formation flying was more than a matter of simple pride for Blenheim pilots. The tactics of the day dictated close formations, in which gunners could offer mutually supporting fire. This pair of aircraft wear the 'XD' codes of No 139 Sqn, based at Bétheniville and Plivot from December 1939 until the end of May 1940, which bore the brunt of much of the heaviest fighting (*via Jerry Scutts*)

even with the protection Spitfires. No 82 Sqn rounded off the day's operations by mounting a three-aircraft attack from which all the Blenheims returned unscathed.

The port of Boulogne fell on 24 May, and it became increasingly clear that an evacuation of the BEF would soon be necessary. No 2 Group's Blenheims would be expected to halt the advancing German Panzers before they could reach the likely evacuation ports of Calais and Dunkirk. As the British Army was pushed closer and closer to the French coast, the Blenheims found that they were operating within the range of friendly fighters almost all of the time, and the provision of escorts kept the loss rate lower than it had been earlier in the month when the German army was bursting over the Franco-Belgian border.

Nos 21, 82 and 110 Sqns mounted missions with six aircraft each early in the day, all returning safely. A single reconnaissance aircraft sent out by No 57 Sqn also returned (albeit crash-landing at Lympne on its return) having been set upon by some 18 Bf 110s. The Blenheim's gunner, Cpl Daley, managed to shoot down the leading Bf 110 into the Channel, after which the remaining Messerschmitts broke off their attacks. Nos XV and 40 Sqns made further attacks (each with six aircraft), all of which returned safely from their missions. Unfortunately, one No.XV Sqn aircraft was lost when it spun in on final approach due to pilot error. In the evening Nos 21, 82 and 107 Sqns mounted further attacks, and again all returned without suffering any casualties.

The importance of fighter escort was demonstrated on 25 May when the Group's solo reconnaissance Blenheim (drawn from No 57 Sqn on this occasion) returned with heavy damage and a badly wounded observer, whilst No 110 Sqn's six protected aircraft returned from their mission without loss. Six aircraft each from Nos XV and 40 Sqns did have a loose escort, but this became separated when they encountered Stukas over the French coast. Deprived of their fighter 'shield', two of the aircraft fell in the face of attacks by nine Bf 109s and very heavy flak. Another

Having flown the first RAF sortie of the war, No 139 Sqn was one of two Blenheim units which moved to France to replace AASF Battle squadrons in December 1939. The unit lost significant numbers of aircraft to flak and fighters in largely fruitless attacks on the advancing enemy armoured columns in the early days of the Battle of France. One of the pilots lost in this wholesale slaughter was Andrew McPherson, who had flown the RAF's first wartime sortie on 3 September 1939. His Blenheim was one of seven squadron aircraft lost on 12 May (*Aerospace Publishing/ Wings of Fame*)

A solitary Bf 109E lands behind a graveyard of wrecked allied aircraft at an unidentified airfield (possibly Le Cateau) in northern France. In the foreground are the burnt out remains of Blenheim IV L9248 of No 57 Sqn, which was posted missing after a raid on Le Cateau on 18 May. Several French Morane Saulnier MS.406s can also be seen behind the Blenheim. The Battle of France bled No 2 Group almost dry, with heavy losses of aircraft (which were relatively easy to replace) and crews (who were not). At the end, as the Germans pushed forward, the RAF even had to abandon Blenheims which were merely damaged, attempting to burn or sabotage them as they retreated (*Aerospace Publishing/Wings of Fame*)

Blenheim (from No 21 Sqn) was lost in the afternoon, but the rest of the aircraft from Nos 21 and 107 Sqns returned safely.

Things could have been very much worse, however. Many had hoped that No 2 Group would be committed entirely to the Battle of France as the second element of the AASF, and it was fortunate that Portal and other senior officers were able to avoid the disaster of sending the squadrons to France, where they would doubtless have been expended in as futile a manner as were the Fairey Battle squadrons. Over the last days of the campaign in France, Blenheim losses declined dramatically, thanks largely to the provision of fighter escorts.

As it was, conserving No 2 Group allowed its Blenheim squadrons to participate in the extraordinary air battle waged to cover the evacuation of the BEF from Dunkirk – codenamed Operation *Dynamo*. Remarkably, and reportedly at Göring's request, Hitler halted the Panzers short of Dunkirk in order to allow the Luftwaffe to bomb the BEF into submission. This was not quite as hare-brained a move as has sometimes been portrayed, since the BEF could have inflicted very heavy casualties on the German army had it tried to take Dunkirk, and might even have held the town for an extended period. History records that the German decision afforded the BEF the opportunity to make a fighting retreat, evacuating 338,000 allied troops by 3 June.

Like other RAF aircraft involved in the fighting, the Blenheims tended to be hitting German units outside the shrinking perimeter, and as such were not seen by the dispirited troops awaiting evacuation, who bitterly asked 'Where are the RAF?' But Britain's total naval supremacy was complemented by the RAF's ability to gain local air superiority, and it was these factors that made the miracle possible. As the evacuation proceeded,

Blenheims pounded the advancing German army units between 26 May and 4 June, when *Dynamo* concluded.

On 26 May a raid by 18 Blenheims of Nos 21, 40 and 82 Sqns returned without loss, as did six aircraft from No 82 Sqn and 18 from Nos 107 and 110, although Nos 53 and 57 each lost a single recce aircraft and No 59 had two downed. The Blenheim IV was proving tough and resilient, however, and many returned to base with severe damage, allowing their crews to fight another day. For example, one of the No 59 Sqn recce aircraft was so severely damaged on 26 May that its pilot ordered the observer to bail out (becoming a PoW), before changing his mind and limping back to base.

27 May started even better, with Nos XV, 21 and 40 Sqns each sending out six aircraft in the morning, all of which returned safely, and with Nos 21 and 82 Sqns repeating the feat in the afternoon – as well as shooting down a Bf 110 in the process. No 53 Sqn was less lucky, for of the three recce aircraft it sent out, one was lost over France and another had to be abandoned over Dover on the return journey'. The final aircraft made it back to base badly damaged. A six-aircraft raid by No 110 Sqn returned without loss, but things went less well for an 18-aircraft effort mounted by Nos 40 and 107 Sqns. Basil Embry, promoted to take command of West Raynham, decided to fly one last mission to show his replacement the ropes, and was duly shot down, his gunner being killed. Embry himself was taken prisoner but managed to escape (after a string of adventures) and eventually returned to England in August. No 107 Sqn lost another aircraft on the same mission.

The situation on 28 May looked very bleak. Belgium capitulated, all available fighters were committed to covering the Dunkirk evacuation and the weather was lousy. But despite the deteriorating situation, things went well for the Blenheim. No 2 Group's units mounted a number of missions, with Nos XV, 40, 82, 107 and 110 Sqns recording no losses, and with No 21 losing a single aircraft. No 59 Sqn also lost a sole Blenheim, although its pilot survived and hitch-hiked back from Dunkirk. Nos 114 and 139 Sqns finally returned to England on 29 May, and raids by Nos XV, 21, 40, 82, 107 and 110 Sqns recorded no losses. One No 21 Sqn aircraft was sufficiently damaged to force a wheels up landing upon its return, but had been responsible for downing a Bf 109. The 30 and 31 May also passed without loss to the Blenheims, despite all the No 2 Group squadrons being active. Spitfires and Hurricanes operating over Dunkirk intervened on a number of occasions to deal with enemy fighters, whilst a Bf 109 was shot down by a Blenheim gunner on the last day of May.

Starting June off on the same note, No 2 Group Blenheims recorded no losses in the first 48 hours of the new month, and after what had been a relatively long period without heavy casualties, most squadrons began returning to the practise of mounting operations with their full complement of 12 aircraft. But the evacuation from Dunkirk was not the end of the Battle of France, though from then on most people realised that the end was both inevitable and impending. The participation of No 2 Group continued much as it had done before Dunkirk, and there were highlights and tragedies.

On 6 June Flt Lt Robert Batt of No 40 Sqn flew an eventful solo recce

mission during which his observer claimed an enemy fighter as a possible – he was awarded the DFC, and both gunner and observer received DFMs. On the same day, his squadron lost five aircraft during a 12 Blenheim attack. On 7 June No 107 Sqn sent three aircraft on a recce. One made a forced landing and was burned by its crew, while Wg Cdr Stokes survived an attack by three Bf 109s, whose enthusiasm was blunted after his gunner despatched one of the fighters.

On 9 June No 107 Sqn lost three aircraft, a fate which befell No 21 two days later, although the latter unit also claimed a pair of Bf 109s during the same operation. Thereafter, UK-based Blenheims continued the fight at a steadily reducing tempo, despite the surrender of the 51st Highland Division on 12 June and the fall of Paris two days later. The Blenheims again participated in covering a British evacuation (this time from French Atlantic ports) on 18 June – No 82 Sqn flew what could be regarded as the last Blenheim operation of the Battle of France on that day.

Missions and losses continued, although these were never again as heavy as they had been in mid-May. France requested an armistice on 17 June and finally surrendered five days later. The Battle of France had cost the RAF 200 Blenheims – 37 from the AASF, 41 from the Air Component and 97 from Bomber Command. The remainder came from Fighter and Coastal Commands.

Everything changed again. Britain now faced the imminent threat of invasion. Suddenly the RAF was the only means by which Britain could conduct offensive operations in Europe, and new responsibilities and duties fell to the Blenheim. Churchill himself remarked that 'The Navy can lose us the war, but only the Air Force can win it . . . The fighters are our salvation, but the bombers alone provide the means of victory'. Everything changed, and nothing changed, for No 2 Group's Blenheims continued to hit targets in France (principally Luftwaffe airfields) and losses continued to mount.

An unidentifiable Blenhem IV of No 114 Sqn is picked over by Wehrmacht troops at Coude, near Vraux, in France in June 1940

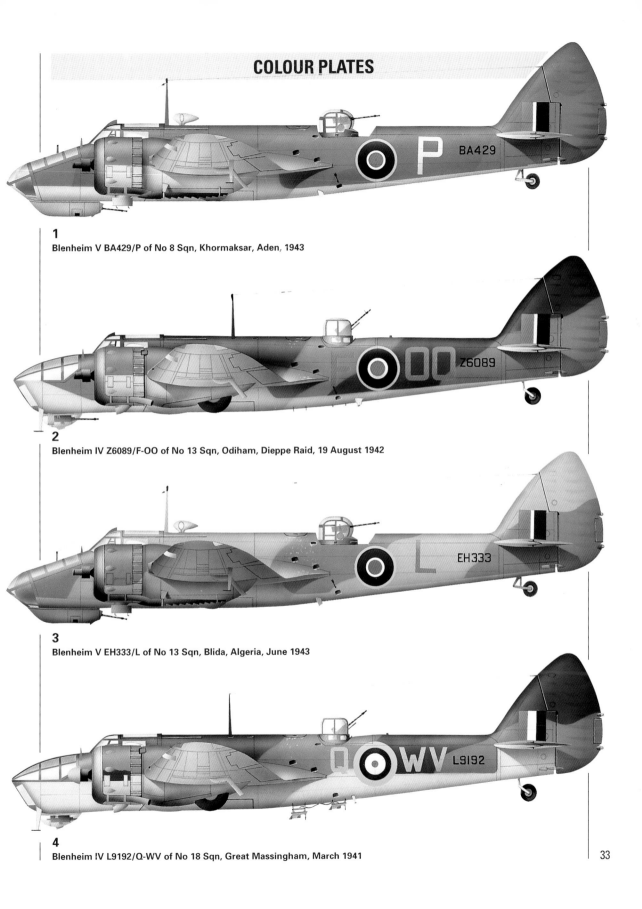

1
Blenheim V BA429/P of No 8 Sqn, Khormaksar, Aden, 1943

2
Blenheim IV Z6089/F-OO of No 13 Sqn, Odiham, Dieppe Raid, 19 August 1942

3
Blenheim V EH333/L of No 13 Sqn, Blida, Algeria, June 1943

4
Blenheim IV L9192/Q-WV of No 18 Sqn, Great Massingham, March 1941

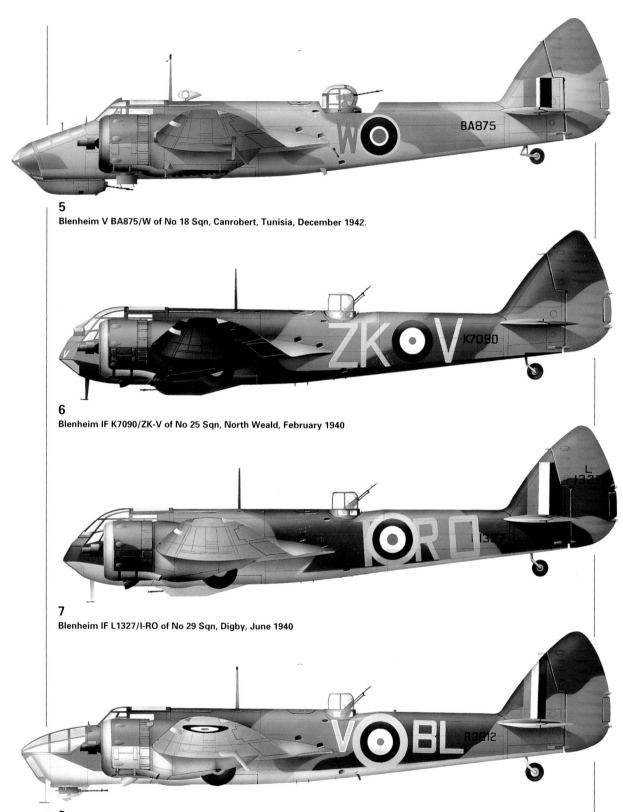

5
Blenheim V BA875/W of No 18 Sqn, Canrobert, Tunisia, December 1942.

6
Blenheim IF K7090/ZK-V of No 25 Sqn, North Weald, February 1940

7
Blenheim IF L1327/I-RO of No 29 Sqn, Digby, June 1940

8
Blenheim IV R3612/V-BL of No 40 Sqn, Wyton, July 1940

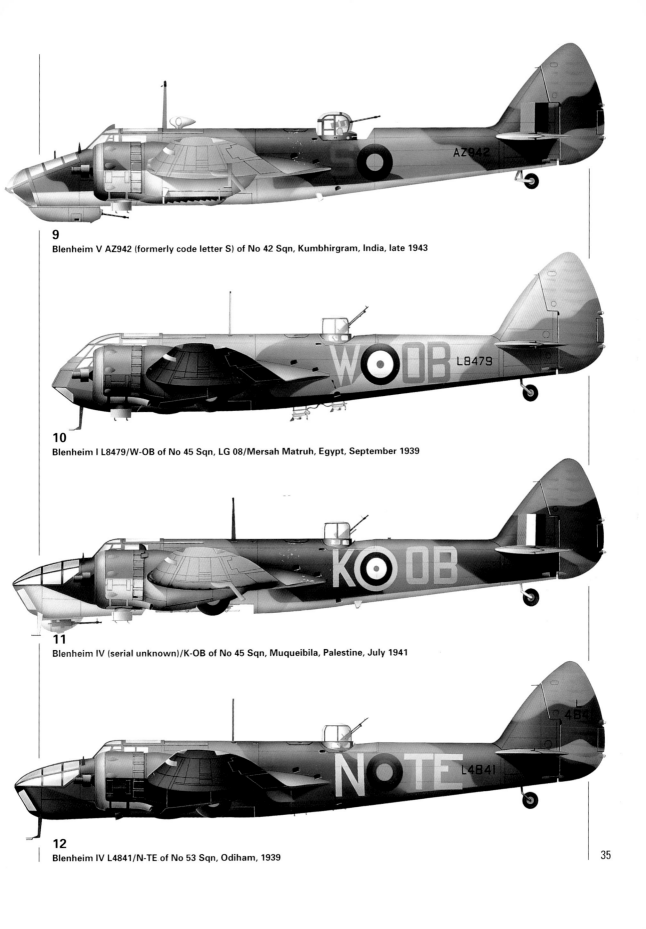

9
Blenheim V AZ942 (formerly code letter S) of No 42 Sqn, Kumbhirgram, India, late 1943

10
Blenheim I L8479/W-OB of No 45 Sqn, LG 08/Mersah Matruh, Egypt, September 1939

11
Blenheim IV (serial unknown)/K-OB of No 45 Sqn, Muqueibila, Palestine, July 1941

12
Blenheim IV L4841/N-TE of No 53 Sqn, Odiham, 1939

35

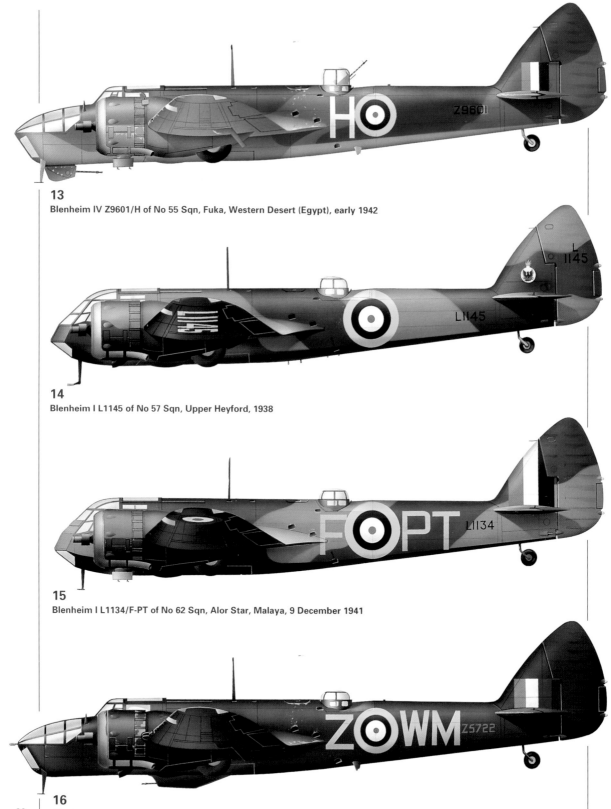

13
Blenheim IV Z9601/H of No 55 Sqn, Fuka, Western Desert (Egypt), early 1942

14
Blenheim I L1145 of No 57 Sqn, Upper Heyford, 1938

15
Blenheim I L1134/F-PT of No 62 Sqn, Alor Star, Malaya, 9 December 1941

16
Blenheim IVF Z5722/Z-WM of No 68 Sqn, High Ercall, 1941

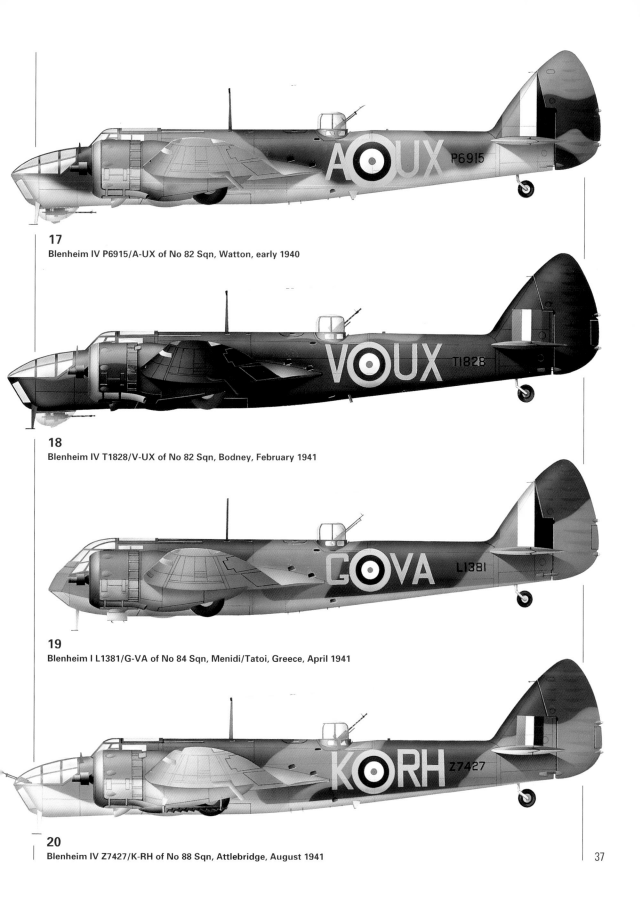

17
Blenheim IV P6915/A-UX of No 82 Sqn, Watton, early 1940

18
Blenheim IV T1828/V-UX of No 82 Sqn, Bodney, February 1941

19
Blenheim I L1381/G-VA of No 84 Sqn, Menidi/Tatoi, Greece, April 1941

20
Blenheim IV Z7427/K-RH of No 88 Sqn, Attlebridge, August 1941

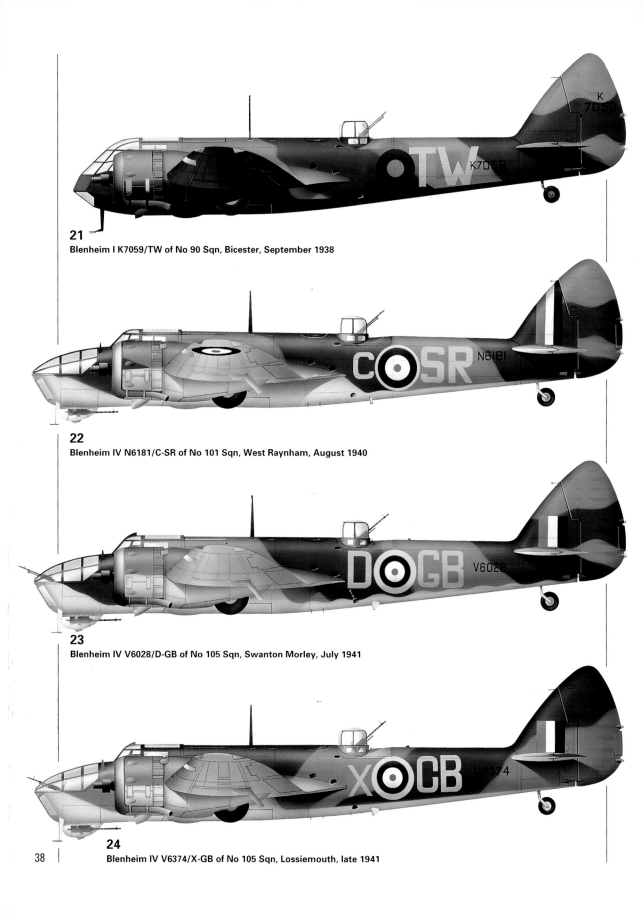

21
Blenheim I K7059/TW of No 90 Sqn, Bicester, September 1938

22
Blenheim IV N6181/C-SR of No 101 Sqn, West Raynham, August 1940

23
Blenheim IV V6028/D-GB of No 105 Sqn, Swanton Morley, July 1941

24
Blenheim IV V6374/X-GB of No 105 Sqn, Lossiemouth, late 1941

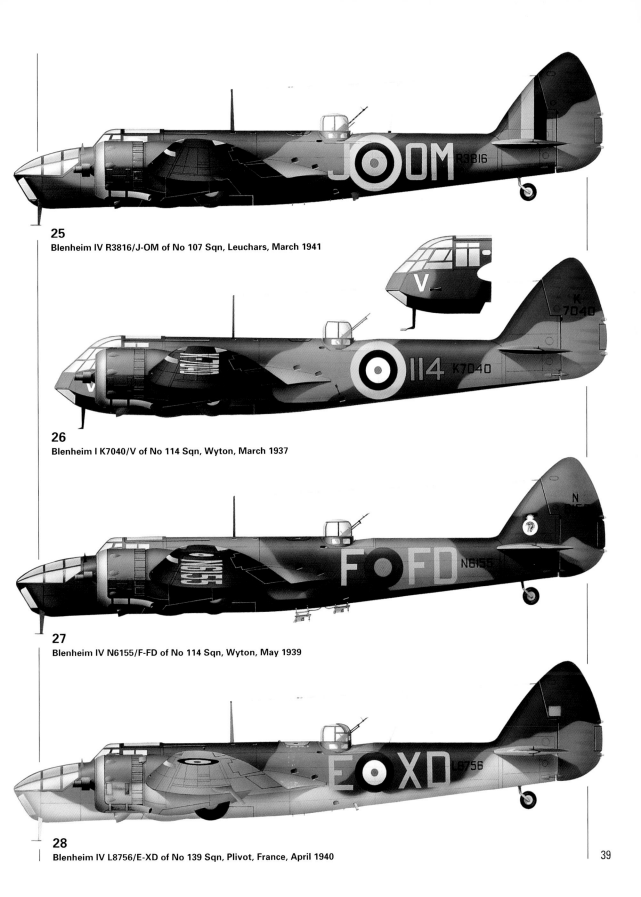

25
Blenheim IV R3816/J-OM of No 107 Sqn, Leuchars, March 1941

26
Blenheim I K7040/V of No 114 Sqn, Wyton, March 1937

27
Blenheim IV N6155/F-FD of No 114 Sqn, Wyton, May 1939

28
Blenheim IV L8756/E-XD of No 139 Sqn, Plivot, France, April 1940

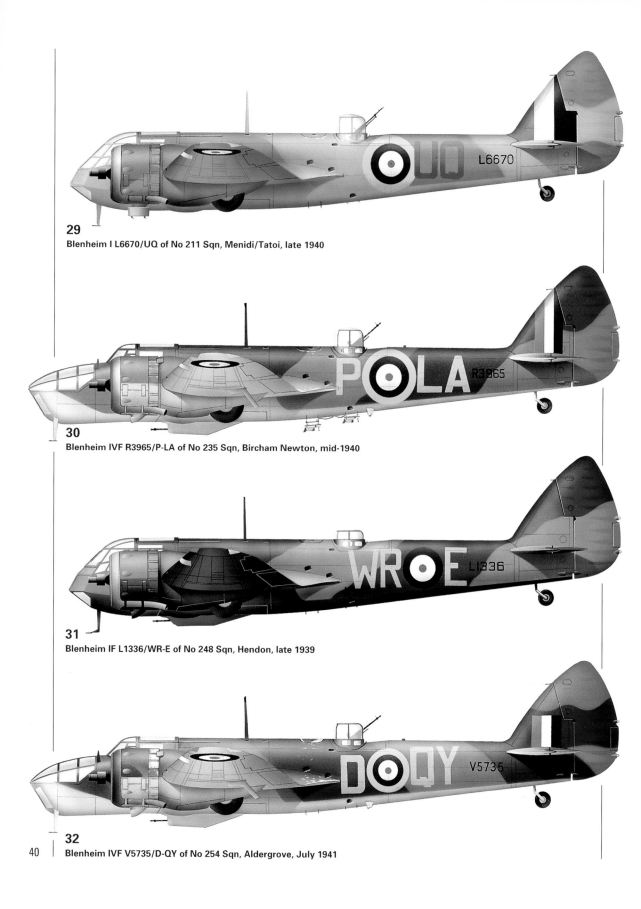

29
Blenheim I L6670/UQ of No 211 Sqn, Menidi/Tatoi, late 1940

30
Blenheim IVF R3965/P-LA of No 235 Sqn, Bircham Newton, mid-1940

31
Blenheim IF L1336/WR-E of No 248 Sqn, Hendon, late 1939

32
Blenheim IVF V5735/D-QY of No 254 Sqn, Aldergrove, July 1941

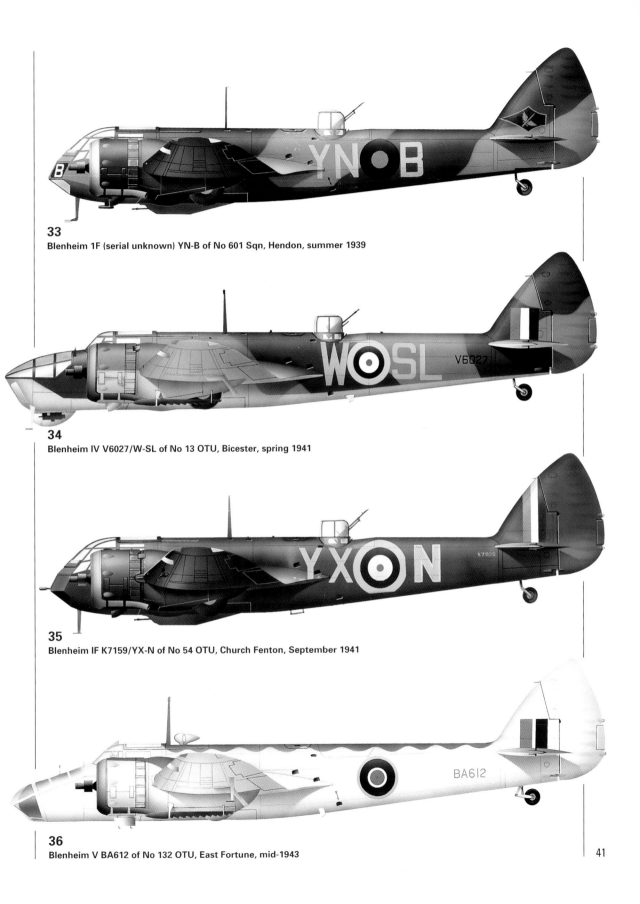

33
Blenheim 1F (serial unknown) YN-B of No 601 Sqn, Hendon, summer 1939

34
Blenheim IV V6027/W-SL of No 13 OTU, Bicester, spring 1941

35
Blenheim IF K7159/YX-N of No 54 OTU, Church Fenton, September 1941

36
Blenheim V BA612 of No 132 OTU, East Fortune, mid-1943

37
Blenheim IVF Z7513/B of No 15 Sqn, SAAF, Cyrenaica, Libya, April 1942

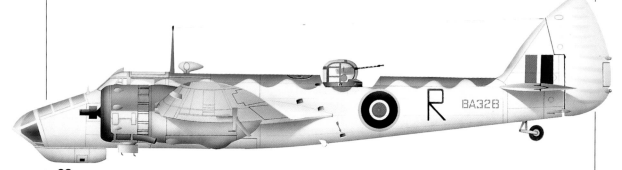

38
Blenheim V BA328/R of No 13 Sqn, Royal Hellenic Air Force, Aden 1943

39
Blenheim IV of *Groupe Lorraine*, Free French Air Force, attached to No 270 Wing, RAF, Fuka, Western Desert (Egypt), November 1941

40
Blenheim V BA849 of *Groupe Bretagne*, Free French Air Force, initially attached to the RAF's No 203 Sqn, Ben Gardane, Tunisia, April 1943

1
Sgt George Keel, Wireless
Operator/Air Gunner with
No 235 Sqn at Thorney Island in
October 1940

2
Sgt S W Lee, wireless
Operator/Air Gunner with
No 113 Sqn at Heliopolis, in Egypt,
in January 1940

3
Sgt Stuart Bastin, Wireless
Operator/Air Gunner with
No 105 Sqn at Swanton Morley in
July 1941

4
Sqn Ldr Hector Lawrence, Officer
Commanding B Flight of No XV
Sqn at Wyton in May 1940

5
Sgt T Inman, pilot with No 82 Sqn
at Watton in March 1941

6
Sgt J W Davies, navigator with No
600 Sqn at Manston in June 1940

ON THE OFFENSIVE

The heavy losses suffered by the Blenheim during the so-called 'Phoney War' and Battle of France continued after France fell, and No 2 Group persevered with mounting attacks which did little for the overall war effort. Squadrons decimated in the ferocious battle for France were reformed, with callow and raw recruits augmenting the few survivors. When one No 18 Sqn observer flew his first operational sortie on 1 August 1940, the trip was his first in a Blenheim, as it was for the air gunner, while the pilot had 38 hours on type! But occasionally such inexperienced crews enjoyed 'beginner's luck'. On this occasion the gunner, a Sgt Bassett, downed the leader of three II./JG 27 Bf 109s which scrambled to attack the lone Blenheim as it bombed Brussels/Evére.

The German fighter unit had been withdrawn to Brussels to recover from heavy losses sustained in the Battle of Britain, but Blenheims shot down two of the unit's Bf 109s in the air, and destroyed three more on the ground, killing the 5th *Staffel*'s *Staffelkapitän*, Hauptmann Albrecht von Ankum-Frank. Airfield attacks were occasionally very successful. On 7 August only two of 29 Blenheims hit their targets, but one of these aircraft dropped a stick of bombs on Haamstede which destroyed two Bf 109s, damaged four more almost beyond repair, killed five airmen, and injured a further 17. The *Staffel* had to be withdrawn from operations for a month to recover from this blow – caused by just one No 82 Sqn Blenheim!

While the threat of invasion remained, Bomber Command was thrown into operations against channel shipping and the invasion barges in their channel ports, while Coastal Command patrolled incessantly, on the lookout for any signs of forthcoming invasion. The former force reduced the 3000-strong fleet by about 12 per cent (including 214 of the 1918 assembled Rhine barges), convincing Hitler that the air situation made invasion too dangerous a gamble. In many respects this was the real Battle of Britain, and certainly it was this little-reported campaign that foiled Germany's proposed Operation *Sea Lion* at least as much as the attrition imposed on the Luftwaffe by Fighter Command. The Führer turned his attention eastwards, while the U-boats intensified their campaign to

Civilian manpower manhandles one of No 13 Sqn's Blenheim IVs, with a steamroller standing by to give the aircraft a tow. The exact circumstances surrounding this scene remain a matter for conjecture, but what is beyond dispute is that No 13 Sqn moved to Odiham on 14 July 1941, becoming part of Army Co-operation Command, which had lost its original Blenheim IV squadrons to No 2 Group and Coastal Command in mid-1940. The Command gained two Blenheim IV squadrons, Nos 13 and 614, as its own tactical bombing force, primarily for anti-invasion duties, with a further unit, No 1416 Flight, operating in the reconnaissance role (*via Peter H T Green*)

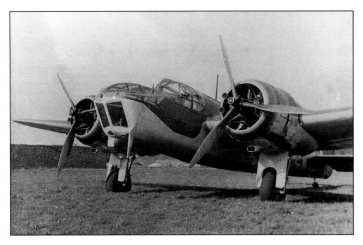

Army Co-operation Command's reconnaissance unit, No 1416 Flight at RAF Benson, gained full squadron status on 17 September 1941, becoming No 140 Sqn. Blenheims continued to fly in the recce role until August 1942, but some remained on the squadron's charge as hacks until August 1943. The squadron conducted many night reconnaissance missions over France, sometimes using photo-flash bombs in what amounted to an operational evaluation (*via Peter H T Green*)

starve Britain into submission. Blenheims then played their part in the war against the U-boat, principally in attacks against submarine pens, shipping and ports.

Not all of Bomber Command's objectives were terribly important in their own right. During early 1941 it participated in 190 'Circuses' – heavily escorted raids whose primary intention was to draw fighters up to do battle with RAF fighters. Some of these were made against land targets, others against shipping. All were very costly, as casually recounted by Laddie Lucas in *Out of the Blue* – 'After the attack on the convoy, during which two out of the three Blenheims were lost – just about par for the course at this time – the striking force turned for home'.

Following the Nazi invasion of the USSR on 22 June, these missions assumed greater importance since they could also be presented as being a way of drawing German attention from the East. But however they were justified, daylight attacks in general, and 'Circuses' in particular, proved costly, and of limited usefulness. During the period between mid June and the end of 1941, RAF fighter pilots claimed 731 aircraft destroyed (most of them on 'Circus' operations), whereas actual Luftwaffe fighter losses were only 103. The RAF lost 123 fighter pilots in six weeks of 'Circus' operations alone. 'Circuses' were equally costly for the Blenheims.

Again and again, whole squadrons would ineffectively attack a heavily defended target and then fail to return, or return in ones or twos. Sorties by individual aircraft or pairs proved no less costly. To make matters worse, No 2 Group was commanded by Air Vice Marshal (AVM) Donald Stevenson, a controversial figure whose 'negative aspect' was summed up by Max Hastings in his book *Bomber Command*;

'He was christened "Butcher", not with the rueful affection with which the name was attached to Harris, but with bitter resentment. An arrogant, ruthless man with no apparent interest in the problems facing his crews, Stevenson seemed to regard No 2 Group's operations solely in the light of his own advancement. "A ship hit is a ship sunk!" he declared emphatically as he compiled wilfully and grossly exaggerated statistics of his Group's achievements. "Churchill wants it!" he declared, incontrovertibly. Stevenson's staff officers, many of whom (like Sam Elworthy and "Paddy" Bandon) were distinguished former Blenheim unit commanders, tried desperately to persuade him to stop the slaughter which they had seen at first hand. During one such discussion, Stevenson even threw an ink bottle at Elworthy, shouting his mantra "Churchill wants it!"'

This was not, in itself, entirely true. In one of his 'Action This Day' minutes to the Chief of the Air Staff, Churchill stated his own position;

'The loss of seven Blenheims out of 17 in the daylight attack on merchant shipping and the docks at Rotterdam is most severe. Such losses might be accepted in attacking *Scharnhorst*, *Gneisenau* or *Tirpitz*, or a southbound Tripoli convoy, because, apart from the damage done, a first-class strategic object is served. But they seem disproportionate to an attack on merchant shipping not engaged in vital supply work. While I greatly admire the bravery of the pilots I do not want them pressed too hard. Easier targets giving a higher damage return compared to casualties may be more often selected.'

This was stern rebuke to No 2 Group's commanders and planners. In a subsequent message to the Blenheim crews, Churchill compared their devotion to the Charge of the Light Brigade, which, he said, was 'eclipsed in brightness by these almost daily deeds of fame'. While the Charge of the Light Brigade was brave, it was also foolish, unnecessary, ill-conceived and counter-productive. Churchill clearly felt the same about the more suicidal Blenheim attacks! Moreover, many of Stevenson's own subordinates considered him to be ruthless and unfeeling, prepared to impose heavy losses on his crews to serve his own ambitions, and considerable doubt was thrown on the accuracy of his shipping claims. Unfortunately, Stevenson was in no mood to hold back, and virtually ignored a directive of 30 June, which effectively passed the anti-shipping campaign to Coastal Command, and he had to be directly ordered to conserve his aircraft and crews in another directive dated 30 August 1941. But when Stevenson was removed from his post in December 1941, it was not in disgrace, and he was sent to command the RAF in Burma.

Some still feel that Churchill's message was disingenuous, and that he was showing a politician's selective amnesia. Certainly, there can be no argument that on 6 March 1941 the Prime Minister had issued a directive to Bomber Command giving the war at sea absolute priority, in response to Hitler's declared intention (stated on 24 February) to use naval blockade to force Britain to her knees within 60 days. To a certain extent it could thus be said that Stevenson had only done what he had been ordered to do with the pitifully weak resources under his control, but a better leader might have questioned his orders, or worked harder to ensure that his men had the tools to carry them out.

From 5 June 1940 Bomber Command directed that No 2 Group's Blenheims would carry out hit-and-run raids aimed specifically at forcing the enemy to deploy his fighters over the widest possible front. A reluctant concession was that such attacks would be carried out only when there was sufficient cloud cover to give some chance of avoiding enemy fighters. It was judged that 7/10ths cover would be sufficient for this purpose. The fine summer weather in 1940 ensured that such attacks were very rare. Unfortunately, when winter eventually set in, there was little improvement, and crews, bored with having to turn back time after time, pressed the limits.

But there were Blenheim operations which were quite literally of vital, war-winning importance. Though no single raid caused conclusive damage to the German battleships 'holed up' in Brest harbour, the nightly bombing by the 'heavies', and occasional pin pricks by the Blenheims, eventually provoked the Kriegsmarine into the reckless gam-

This No 40 Sqn Blenheim was reported missing near Ostend on 9 September 1940. The aircraft has its individual code letter thinly outlined in white. The squadron re-equipped with Blenheims in December 1939, having been a Battle unit within the AASF. It began operations during the Battle of France and fought on without much respite until it re-equipped with Wellingtons in November 1940 (*via Phil Jarrett*)

No 101 Sqn flies past, the bomb bays of its Blenheim IVs open – the leading aircraft in this formation was shot down by flak off Boulogne on 3 May 1941. No 101 Sqn initially served as a training unit, and did not make its first bombing raid until 4 July 1940. It began to re-equip with Wellingtons from April 1941, discarding the last of its Blenheims the following month. The conversion of a number of No 2 Group Blenheim squadrons made aircraft available for additional Coastal Command Blenheim units

(*Charles E Brown via Phil Jarrett*)

ble of the 'Channel Dash'. The RAF failed to sink the ships as they ran for home, but that was far from the end of the story. *Scharnhorst*, *Gneisenau* and *Prinz Eugen* got through, but *Gneisenau* was so badly damaged by mines and subsequent air raids on Kiel that she spent the rest of the war in dry dock, and *Prinz Eugen* was repaired too late to play an active part in the war. *Scharnhorst*, meanwhile, was forced into a nomadic existence avoiding allied bombers and warships, until her luck ran out in the North Sea. None of these ships were able to break out into the Atlantic to wreak havoc on allied convoys largely as a result of Bomber Command's raids.

Less spectacularly, but no less importantly, from 15 June 1940 the Blenheims of No 2 Group began to raid German aerodromes in occupied France, escorted by Hurricanes when the targets were near enough to the coast – these operations continued throughout the Blenheim's operational career. Airfield attacks had obvious potential risks, yet remarkably the loss rate remained low . . . usually. There were of course exceptions, one such involving Nos 21 and 57 Sqns, who lost six aircraft attacking Stavangar/Sola on 9 July, while No 107 lost five of six aircraft attacking Amiens the next day. Worse was to befall the unlucky No 82 Sqn (which had previously lost 11 of 12 Blenheims over Gembloux in May 1940).

On 13 August 1940 Wg Cdr de Virac Lart led 12 aircraft across the North Sea to raid Aalborg, but crossed the Danish coast 30 miles too far south, giving an extended overland run during which all the aircraft were shot down, killing 20 of the 33 crew. Five of the Blenheims fell to flak, the others being claimed by the Bf 109s of JG 77 – one aircraft which had turned back early with fuel problems was the sole survivor. No 82 Sqn's run of bad luck continued, losing five of nine aircraft on the night of 3 December 1940. The Blenheim could occasionally surprise its critics, however. On 30 October 1940, for example, Sqn Ldr Little of No 40 Sqn

happened upon a Ju 88 during a training flight over East Anglia and promptly shot the intruder down at Stuntney, near Ely.

On 17 August 1940 the Medium (bomber) Expansion Re-equipment Policy Committee met to discuss the continued use of the Blenheim by Bomber Command, partly in reaction to a recent statement by the Command's AOC-in-C that 'it would pay us to roll up some if not all the Blenheim squadrons and convert them into heavy bomber squadrons . . . although they are doing a certain amount of useful work . . . far better value would be obtained by applying the personnel in a smaller number of heavy bomber squadrons'. But other factors were at work, including an Air Council promise to the Army that 250 aircraft (15 units) would be available for army support. In the end, it was decided that Bomber Command could not be weakened for the time that such a re-equipment would take, but it does indicate how dissatisfied many were with the Blenheim even as early as the summer of 1940.

Generally, the aircraft continued to operate by day, since good weather conditions and moonlight were required for accurate navigation in the precision bombing role, yet enough cloud cover had to be available to hide from enemy fighters. No 2 Group continued to try nocturnal operations, but the lack of suitable nights and the aircraft's very limited bombload made these operations of marginal usefulness. Nos XV, 40, 57, 101 and 218 Sqns re-equipped with Wellingtons in November 1940, considerably reducing the number of Blenheims available for the continuing daylight offensive. Despite this, Blenheims played a significant role in Bomber Command's early large-scale night raids, nine of the remaining aircraft participating in Bomber Command's first mass raid (against Manheim on 16-17 December 1940).

Blenheims also played a significant role in the attack on Wilhelmshaven town on 15 January 1941 (in reprisal for the Luftwaffe's attack on Plymouth), which was then the RAF's biggest raid of the war. Sometimes even airfield attacks were made at night, with the more adventurous crews timing their attacks to coincide with the return of German bombers to their bases. Blenheim pilots would occasionally fire off the German colours of the day (if these could be ascertained) and then enter the circuit with their navigation lights on. Sometimes Blenheims would simply bomb the flarepath, putting it out of action and forcing enemy aircraft to divert or to run out of fuel waiting for the flarepath to reappear.

The 'Blenheim boys' were probably less vulnerable than most, since

No 40 Sqn at RAF Wyton put up this neat formation in early March 1941. Vics of three aircraft were supposed to provide the best mutual fire against attacking fighters, but it was already apparent that cloud cover and ultra-low level tactics made far better sense (*via Aeroplane*)

A Blenheim crew disembark after a sortie during the summer of 1941. The observer is in shirtsleeve order, but the gunner still wears his fleece-lined Irvine jacket. The installation of twin-Brownings in the dorsal turret effectively doubled the defensive firepower of the Blenheim, since the undernose gun was of limited accuracy and usefulness. It has been a long trip home, and relief at having survived is tempered by the knowledge that tomorrow, or the next day, they will have to run the gauntlet again (*via Phil Jarrett*)

they flew an aircraft that was relatively small and difficult to see, and one with sufficient agility to be flown at very low level with some degree of confidence. And low level really meant low level. By day, the Blenheims navigated by church steeples, and typically flew at about 40 ft, and even lower over open fields. Blenheims regularly returned to base trailing bits of telegraph wire, or with clumps of vegetation stuck in the tailwheel. One crew even claimed to have killed a mounted soldier during a low level turn, a claim apparently borne out by the sinister dent in the leading edge of the port wingtip.

Much of the Blenheim's work with No 2 Group was against Axis shipping, these Bomber Command aircraft augmenting the Blenheims of Coastal Command. After Churchill's 6 March 1941 directive, four of No 2 Group's Blenheim units (Nos 18, 21, 107 and 139) were specifically tasked with halting enemy convoys by day, and other squadrons also played a part in the campaign. The coast of occupied Europe from Bergen to Oslo and from northern Denmark to Bordeaux was divided into 19 'Beats', which would be patrolled by Blenheims on the lookout for ships.

But anti-shipping operations were dreadfully costly, since such convoys often had fighter cover, sometimes flew barrage balloons and always included heavily-armed flak ships. Operations against enemy ships were collectively known as 'Roadsteads', and such missions became more common from March 1941, when Churchill promulgated them as being a 'top priority'. Between mid-March and mid-April 1941 nine Blenheims were lost on 'Roadsteads', and 36 more fell in the three months from mid-April to mid-June (these representing 12 per cent of aircraft despatched). During this period, No 2 Group mounted 297 anti-ship attacks, losing 36 aircraft, while Coastal Command despatched 143 aircraft, losing 52 of them. August saw the loss rate soar to 30 per cent, with 23 of the 77 aircraft despatched failing to return. Only 13 Blenheims engaged in other daylight operations were lost in the same month.

From 31 March 1941, aircraft involved in the anti-shipping campaign began mounting so-called 'fringe attacks', penetrating in from the coast to attack land targets on a 'hit-and-run' basis. This forced the Germans to further extend their flak and fighter defences, and incidentally provided a much-needed morale booster for the people of occupied Europe. The sight of a formation of Blenheims thundering inland at very low level in broad daylight was an immediate reminder that the war against Hitler was continuing. From April 1941 detachments (usually of a squadron of Blenheims) were maintained at Manston for 'Channel Stop' operations which aimed to completely close the English channel to enemy shipping at Dover. The aircraft (usually operating in threes) were heavily escorted by Spitfires and Hurricanes, but losses were extremely heavy, and many units had to be withdrawn after only a couple of weeks.

Over water, the Blenheims operated at very low level, attacking from mast height and making skidding turns to avoid banking too much. This ran the risk of exposing the aircraft's belly to hostile fire, and made it more likely that a wingtip could hit the water. Low flying sometimes prevented enemy flak ships from firing, for fear of hitting ships behind the wave skimming Blenheim. Ted Sismore, a Blenheim navigator, recalled that;

'We did low-level anti-shipping (attacks) anywhere down the coast from Denmark to Western France . . . We did hit the sea one night, off the entrance to the Kiel Canal. It was a misty, foggy, night. We were looking for a ship that was reputedly in the Canal, and as we turned away from this very flat country we hit the water and bounced off. It transpired later that we'd lost one propellor blade and we'd bent back the other five.'

Blenheim operations were full of examples of such casual heroism, though the success rate of the operations against enemy ships was low. Natural justice might have demanded that a VC be awarded to someone participating in these suicidal missions, and there were many episodes of extreme bravery which would have warranted such an award. A No 2 Group Blenheim anti-shipping VC would have gone some way towards recognising the sacrifices made by all those who participated, but unfortunately, such recognition was not given.

At least one Blenheim pilot involved in the campaign was recommended for a VC, however, by No 2 Group's AOC, the controversial AVM Stevenson. This pilot was Plt Off Edgar Phillips of No 139 Sqn, who was fatally wounded during his 43rd operational sortie. Phillips retained control long enough for his gunner and navigator to lift him from his seat, and for the navigator to then take the controls and fly home. Despite his wounds, Phillips (who had been lapsing in and out of consciousness throughout the return flight) took control for the landing, but died in hospital that night. The navigator got a DFC for his part in the episode, but the gunner (recommended for a DFM) received no award.

At the time it was thought that 41 enemy ships were sunk in the anti-shipping campaign (as a result of 698 attacks), but postwar research revealed that only 29 vessels were sunk and 21 damaged in 'Roadstead' operations. Some 123 Blenheims were lost in return. Efforts were made to reduce losses, including the mounting of moonlit anti-shipping strikes from the end of May 1941, and the beginning of anti-shipping 'beats' with fighter escort from 18 August – this was a development Stevenson had pressed for since April.

Flt Sgt Shackleton of No 114 Sqn described a typical mission, mounted in September 1941, looking for a large enemy merchant ship off the French coast;

'It was our first show. The wing commander took the middle "beat" and we were on his left. We didn't see the big ship, but after turning to port and flying for the prescribed length of time along our "beat" we saw a convoy just on the horizon. At that moment too, we saw a big ship,

A Blenheim IV of No 105 Sqn lands on Swanton Morley's new concrete runway in March 1941. With such a runway, the Norfolk bomber base could support operations by the very much more capable Boston. No 105 Sqn was one of the more successful Blenheim units, under the leadership of Arnold Christian and his replacement, Australian 'Hughie' Edwards (*via Jerry Scutts*)

Although of poor quality, this rare line-up shot of No 21 Sqn aircraft includes Blenheims in both standard day bomber camouflage and in temporary night finish. These aircraft were photographed in April 1941 during a detachment to RAF Chivenor (*via Andrew Thomas*)

probably the one we were after, but as we were in the wrong position to attack her we went on for the convoy. I chose a ship of, I judged, between 3000 and 4000 tons. It had a flak ship to starboard.

'I cracked on full boost and went for the merchant ship. Both it and the flak ship opened up. The air in front became full of black puffs with red sparks crossing them. That was tracer and it all looked as though it was going straight towards us, only to turn away and hit the water, where the splashes reminded me of those made by children when they throw handfuls of pebbles into a pond. Flak looks like that when you're coming in low. It was my first attack on a ship, and I pulled up a bit late and only just missed the mast! We then found that we were rushing straight over a destroyer, and that an Me 109 was 400 ft above us! He fired a few short bursts and then went off. The wing commander must have got his ship, because we saw a big one with its decks awash just after our attack.'

On 3 November 1941, at a conference of No 2 Group Station Commanders, the AOC (Stevenson) finally announced the end of the terribly costly anti-shipping campaign, though in fact it continued (albeit at a less intense pace) until his replacement (AVM Lees) took command of the Group. Stevenson was actually directly ordered to conserve his crews on 30 August, when the continued daylight maritime operations were dismissively described as 'wastage'. Not that HQ Bomber Command was motivated by altruism – the Blenheims and their crews were urgently needed elsewhere, most notably on Malta and in the Mediterranean. Thus Bomber Command's No 2 Group was released from anti-ship operations in November-December 1941, officially ending on 25 November, although Coastal Command Blenheims fought on. Indeed, the latter had actually formed two new Blenheim squadrons a year earlier when five Bomber Command Blenheim units had converted to the Wellington. In fact Blenheims of No 2 Group would occasionally be called on to attack shipping in months to come, most notably during the 'Channel Dash'.

It was officially estimated that between March and October 1941, No 2 Group had successfully attacked 590 ships (1,303,000 tons), resulting

Blenheim IV V6240 of No 21 Sqn failed to return from a mission on 12 July 1941 (*via Phil Jarrett*)

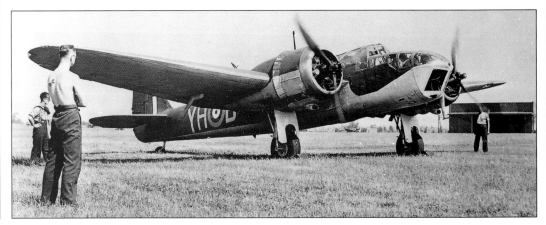

in the sinking of, or serious damage to, 107 vessels (totalling 355,000 tons). In fact, postwar research suggests that the true figure was rather lower, with 29 ships (29,836 tons) sunk and another 21 (43,715) damaged. These losses were negligible, and had little effect on the total tonnage of imports going through Rotterdam, which was the worst affected of the German-controlled ports, although on 13 November 1941, GrossAdmiral Erich Raeder, C-in-C of the German navy, reflected that 'Despite the very difficult circumstances, it has been possible to keep open the shipping lanes . . . Losses have reached the limit of what we can sustain'.

Weighed up against this modest level of success, the campaign against North Sea and Channel shipping had cost No 2 Group the lives of its most experienced crews, none of whom had managed to

complete 50 operations or 200 hours. The virtual impossibility of completing a tour had led to plummeting morale and a rising cynicism. Few were not aware that the campaign had been unimaginitively and inflexibly planned and undertaken, with no attempt to vary tactics or to reduce the level of casualties. 170 Blenheims were lost during the same period, excluding aircraft destroyed during detachments to Malta or on 'Circus' raids against land targets. It was fortunate that the German fighter force in the West had been reduced to skeleton strength throughout 1941 in preparation for, and then participation in, the invasion of the USSR.

The first Blenheim pilot to win the VC was Australian Wg Cdr Hugh Idwal Edwards, injured in a pre-war flying accident which paralysed his right leg below the knee, and who had already won a DFC for a daring low-level raid against shipping during June 1941. Although Edwards flew in the anti-shipping strikes of 1941, his VC was awarded for an overland operation. He flew with No 139 Sqn from Horsham St Faith before taking command of No 105 at nearby Swanton Morley. Edwards was not universally popular at first, not least because his leadership style was very different to that of his equally successful predecessor, Arnold Christian.

Christian had been a superb pilot (a former CFS A1 category instructor) who led by example, yet who combined good humour with a genuine compassion for his crews, who felt themselves to be 'his boys'. Edwards was more of a Martinet, and was aloof, ambitious and strict. He wasn't interested in joining in or making friends. A relatively unimaginative man, Edwards genuinely believed that he could lead a box of six Blenheims over the Ruhr in daylight, and that the combined firepower from the twelve turret-mounted 0.303-in guns would render the forma-

This Blenheim IV was caught by the strike camera of another aircraft in the formation during March 1941. By this stage of the war, Blenheims were usually operating at much lower altitude than this, and would seldom fly in such bright and cloudless conditions (*via Aeroplane*)

On 7 July 1941 five Blenheims from No 105 Sqn and six from No 139 found and attacked a convoy off the Hague. Three of the Blenheims were shot down, and only one of the enemy ships (the *Delaware*) was hit. Eight bombs struck the vessel, but only two exploded, killing two of its crew but causing little damage (*via Bruce Robertson*)

tion safe against fighters, as long as formation discipline was tight enough. It was hardly surprising that Edwards quickly gained a reputation with No 105 Sqn for daring low level strikes. The *London Gazette* printed his VC citation (here reproduced with additions and clarifications in italics);

'Wing Commander Edwards, although handicapped by physical disability resulting from a flying accident, has repeatedly displayed gallantry of the highest order in pressing home bombing attacks from very low heights against strongly defended objectives.

'On 4th July, 1941, he led an important attack (*by 15 Blenheims, six of them drawn from No 107 Sqn at Great Massingham*) on the Fort of Bremen, one of the most heavily defended towns in Germany.'

The citation did not mention that this was the third attempt by No 2 Group to hit Bremen. On 28 June the raid was led by Wg Cdr Lawrence Petley of No 107 Sqn, who abandoned the attack when he lost the element of surprise, being sighted by destroyers on the run-in. AVM Stevenson was livid, caring less for his crews than for his own reputation. A terse phone call between Petley and Stevenson resulted in the former shouting 'If that's the way you feel, we'll go back and do it this afternoon!' Fortunately this did not happen.

For the second attempt, Petley and No 107 Sqn were relegated to the diversionary attack on the fighter airfield on the island of Sylt, while Nos 21 and 105 were to strike at Bremen itself. By this time most of No 107 Sqn's aircraft had been modified locally with extra defensive armament – a rearward-firing 0.303-in Browning in each engine nacelle and a rearward-firing Vickers K-gun under the rear fuselage. The second attempt to attack Bremen was abandoned because the formation broke up in a thick blanket of fog and low cloud over the North Sea, although a handful of

crews pressed on and attacked a number of targets. Plt Off Waples of No 21 Sqn attacked Bremen itself, and evaded attacks by two Bf 110s before being jumped by three Bf 109s, one of which was shot down. Waples returned to make a crash landing at Manston.

The third attempt was postponed twice from 2 July to 4 July, with No 226 Sqn making a diversionary attack on the seaplane base at Norderney, and with the main force provided by Nos 105 and 107 Sqns, which despatched nine and six aircraft respectively. Bremen itself had been 'softened up' by Wellingtons the previous night. Three of the No 107 Sqn Blenheims were forced to turn back, but the remaining 12 pressed on.

'This attack had to be made in daylight and there were no clouds to afford concealment. During the approach to the German Coast several enemy ships were sighted, and Wing Commander Edwards knew that his aircraft would be reported and that the defences would be in a state of readiness. Undaunted by this misfortune (*and perhaps fearful of AVM Stevenson's likely reaction to another abort!*), he brought his formation 50 miles overland to the target, flying at a height of little more than 50 ft, passing through a formidable balloon barrage.'

The formation crossed the coast just south of Cuxhaven, then turned south to Bremen. The weather was fine and clear, a gift to the defenders.

'On reaching Bremen he was met with a hail of fire, all his aircraft being hit, and four of them (*two from each of the participating squadrons, including the aircraft flown by No 107 Sqn's OC*) being destroyed.'

Edwards' tactics were to attack in a widespread formation, with the aircraft about 200 yards apart, across a wide front, forcing the defences to divide their efforts. Bremen was surrounded by a balloon barrage 'close-hauled' to 500 ft, within which was a ring of heavy AAA positions, 'Nevertheless he made a most successful attack, and then with the greatest skill and coolness withdrew the surviving aircraft without further loss'.

The aircraft attacked targets of opportunity at ultra low level., ducking below high tension lines rather than going over them. Even HE bombs were dropped from 50 ft, sending columns of debris hundreds of feet into the air. After the aircraft had dropped their bombs, they machine gunned gun emplacements, barracks and railway yards, all at rooftop height. Several aircraft returned to base with clear evidence of having hit telephone

A Blenheim IV lands at Swanton Morley during late 1941, with a Boston of No 88 Squadron visible in the background. Interestingly, despite the date, this Blenheim has no undernose gun fitted, instead relying on a free-mounted 'scare-gun'

Blenheim crews of No 114 Sqn pose at Odiham on 13 February 1942 – the morning after the 'Channel Dash' operation against the German capital ships *Scharnhorst*, *Gneisenau* and *Prinz Eugen*. The Earl of Bandon can be seen in the great coat (*via Phil Jarrett*)

wires, at least one returning with a length wrapped around its tail-wheel. Edwards circled Bremen after dropping his bombs, strafing a stationary flak train (which returned his fire) before running out over Bremerhaven. He returned with the port wingtip and aileron missing, a cannon shell in the radio, and bullet holes in the belly – his aircraft had taken 20 hits over the target alone. He also brought home an injured gunner, Sgt G Quinn, who had to be extracted from his turret by crane! Operation *Wreckage* was a great success, and represented the Blenheim's finest hour, although losses were heavy, with four aircraft downed.

On 12 August 1941 a force of 54 Blenheims from Nos 18, 21, 82, 107, 114, 139 and No 226 Sqns, escorted part of the way by hundreds of fighters, including long-range Whirlwinds, attacked power stations in the towns of Knapsack and Quadrath. The attacks were pressed home at very low altitude, and caused significant damage. Perhaps even more significantly, the force of 54 Blenheims thundered over Holland at roof-top height – a most welcome sight for the beleaguered Dutch! But the towns were well defended and there was heavy AAA over the target and on the run-in, and the Blenheims had to contend with fighters on the way home.

Wartime HMSO propaganda booklet *Bomber Command Continues* mentioned the raid in detail, quoting an unnamed Blenheim pilot whose aircraft was hit ten times in the port wing before reaching the target;

'We could see the twelve chimneys - a row of four and a parallel row of eight - standing dark against the sky. We climbed to attack because the smoke and flames, some of which were licking to 50 ft high, were too thick for us to go through and bomb accurately. I could see my observer's elbow as he pulled back the release lever, and as he called bombs gone I did a steep turn over a belt of trees down into a quarry to escape the flak.

Coming out of the quarry we ran into fighters. It must have been then that the rear gunner was hit, because we heard no more from him. I went into nine boost for more evasive action, and a bullet came in behind my head and broke the perspex, while another smacked into the armour plating at my back and gave me a slight jerk. I gave the observer a shell bandage and he managed to creep back to the rear gunner. We came back through a gap between two storms and made for base. Our undercarriage had been put out of commission, so the observer had to crawl back once more to hold down the gunner while we made a successful belly landing.'

These attacks were the deepest penetrations of Germany to date, and 12 aircraft failed to return.

On 16 July 1941 Nos 18, 21, 105, 139 and 226 Sqns (with cover from two squadrons of Spitfires) launched 37 aircraft against Rotterdam docks. The first wave consisted of 12 aircraft from No 21 Sqn and six from No 226, and was led by Wg Cdr Peter Webster, who won the DSO for his part in the raid. The second wave, led by Wg Cdr A Partridge, included eight aircraft from his No 18 Sqn, four from No 139 and seven from No 105, one of the latter acting as a marker for the fighter escort. Three Blenheims were lost but 17 ships were claimed destroyed or damaged, and considerable damage was inflicted on the docks' infrastructure. A follow up mission by 17 Blenheims on 28 August went less well, resulting in the loss of seven aircraft (plus some escorting fighters) and prompting Churchill's famous minute to Portal.

A rather different sortie was conducted on 19 August when R3843 of No 18 Sqn flew to its target via the German fighter aerodrome at St Omer, where it dropped a case containing a spare leg for the recently captured Douglas Bader, who had been shot down ten days earlier.

On 27 December 1941, 13 Blenheims drawn from No 114 Sqn and six from No 110 Sqn mounted a daring attack against the Luftwaffe airfield at Herdla, in Norway, to prevent interference in a combined services (commando) raid against Vaagso. The mission was led by the OC No 114 Sqn, Wg Cdr Pollard, with his navigator, Flt Lt Brancker. This had to be timed to the split second, with the Blenheims hitting the airfield's wooden runways within one minute of noon, after a 300-mile over-water flight at very low level. A Bf 109 was able to start its take-off run during

Two-number codes identify this Blenheim as belonging to a training unit (perhaps an OTU), although the wartime caption is not specific as to which unit the aircraft belonged to. But whatever the unit, this represents the textbook height for a low level attack – below this and the Blenheim would be too close to the blast, and in danger of damage from the bomb's shock wave, or even from shrapnel (*via Aeroplane*)

Temporary black camouflage on Blenheim undersides was routine from the early days of the war, and nightfighters went 'all-black' from November 1940. But only a handful of bomber Blenheims (like this No 114 Sqn Mk IV at West Raynham in 1942) were ever painted black over-all, even when the last couple of squadrons switched almost entirely to night intruder work against Luftwaffe airfields (*via Andrew Thomas*)

With so many Blenheims being shot down or force-landed over enemy territory, and with large numbers of aircraft having been abandoned in France and Greece, it was only a matter of time before the Luftwaffe had an airworthy Blenheim to evaluate (*via Aeroplane*)

the attack, but fell into a bomb crater. Five aircraft were lost, but the mission was an outstanding success. Blenheims from No 404 Sqn provided long range fighter cover, and mounted a diversionary attack. Wg Cdr Pollard described the mission;

'The first thing we saw were the mountains of Norway some 70 miles away. It seemed an age before we reached them, but when we got close to the shore we turned to port and flew about ten miles northward some three miles from the coastline. We were looking for the narrow opening into a fjord. For some little time I was confused, but my navigator was never in doubt. There was a man, a girl and a child on the edge of the fjord, waving to us, just as we turned in to attack the aerodrome. I remember saying to myself "For God's Sake get under cover – the shooting is going to start", and indeed it did so as we swept in. The German anti-aircraft defences opened up, but they had been taken by surprise and for a very important minute we were not shot at. During that minute each section of Blenheims made for the section of runway assigned to it beforehand. The bombs were dropped "according to plan". I saw them dig into the snow where they stuck for a moment like darts in a board before going off.

'There were a number of Me 109s with their propellors turning, about to take-off. They failed to do so, and one which was on the move down the runway fell into a crater which appeared just in front of it. We had some casualties. In my formation the port aircraft was hit, reared up over me, nearly collided with me, then sheered off and collided with another Blenheim. The ground staff of the enemy aircraft were all around the runways. Most of them were killed by the bombs or the continuous barrage of machine gun fire put up by our air gunners. We certainly wrote off a great many Germans. On the way back it was wizard to see the Norwegians waving to us. I felt I wanted to stop and pick them all up and take them back to Scotland.'

The Blenheim faded from the scene in Western Europe quickly during 1942, with No 2 Group's squadrons re-equipping with Douglas Bostons, North American Mitchells and Lockheed Venturas. The Boston entered service with No 88 Sqn at the end of 1941, the Ventura with No 21 Sqn in May 1942 and the Mitchell followed (with No 98 Sqn) in October 1942. The eight long-term Blenheim squadrons reduced further to only three units by the end of February 1942. But Blenheims continued to play their part until the end. In February 1942, for example, No 114 Sqn was the RAF's only home-based night intruder unit, while No 18 Sqn completed night-intruder training in April and resumed operations. On the night of 30/31 May, for example, Blenheims

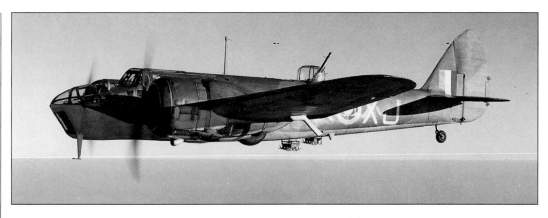

The Blenheim OTUs, No 13 at Bicester and No 17 at Upwood, used a variety of aircraft. This No 13 OTU Blenheim IV has no undernose weapon, and has the original single gun dorsal turret – adequate for training, but long augmented by extra armament on frontline squadrons (*via Phil Jarrett*)

No 13 OTU applied the codes 'FV' and 'XJ' to its Blenheims. This aircraft represents the final Blenheim IV configuration, with twin undernose guns in an articulated turret, and with the later twin-gun dorsal turret (*via Phil Jarrett*)

from Nos 18 and 114 Sqns, and from Army Co-operation Command's Nos 13 and 614 Sqns, were among the aircraft despatched on Bomber Command's first 1000 Bomber Raid, against Köln, although the Blenheims supported the raid by attacking nightfighter bases and other targets, rather than the city itself. The Army Co-operation Command squadrons had respectively deployed to Wattisham and West Raynham – Bomber Command airfields.

According to some sources Blenheims made their last sorties in Western Europe on 17/18 August when No 18 Sqn sent its Blenheim IVs as night intruders against a number of German fighter airfields. In fact this was only the last Blenheim mission over Germany, for the bomber was still operational with Coastal Command and Blenheims participated in the Dieppe Raid on 19 August 1942, when they were engaged in smoke-laying to cover the landings. The Blenheim then disappeared from No 2 Group, although there were nasty moments for some when rumours suggested that the type might be brought back. This was supposedly because the Boston was unable to conduct anti-shipping toss-bombing attacks, and had to deliver its bombs in level flight. Plans to replace Blenheim IVs with the improved Mk V within No 2 Group came to nothing, however, and responsibility for the anti-shipping campaign passed to Coastal Command, whose new Beaufighters were far better suited to the role.

The Blenheim V was originally designed to meet a requirement for an army support aircraft, and was originally known as the Bisley, a name which was officially abandoned, yet which stuck to the aircraft even when

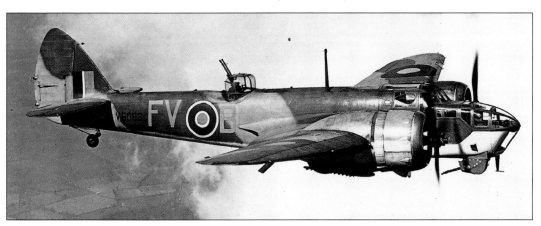

This No 13 OTU aircraft (V6083) was built by Rootes Securities at Speke. The aircraft has cable cutters on the leading edge of the wing, and a camera gun bracket above the undernose turret on the starboard side of the nose (*via Aeroplane*)

This Blenheim IV (V5382) was the subject of a series of photographs taken by the Aeroplane and Armament Experimental Establishment on 8 September 1941 on behalf of the Ministry of Aircraft Production. The aircraft is in typical early configuration, with single gun dorsal turret and early underfuselage turret. It bears traces of heavy usage, with scuffed paint surrounding the hatches, along the trailing edge of the port wing and around the engine cowlings. Interestingly, the demarcation between the upper and lower surface colours is scalloped – never a common feature (*Aerospace Publishing/Wings of Fame*)

it entered service. The first two prototypes were built with a solid nose housing four Browning machine guns, and the new variant was much better armoured than the Blenheim IV. This made the Mk V significantly heavier than the Mk IV, while its Mercury 25 or 30 engines were only 30 hp more powerful than the Mercury XVs of the earlier Blenheim. This led to the Mk V being somewhat underpowered in service. By the time production began, fighter-bombers were undertaking the close support role, and the type was built with an alternative half-glazed nose as a conventional light day bomber, although the new version retained its improved crew protection. The new nose was entirely asymmetric, with an extended glazed navigator's section to port, and with an undernose fairing and twin rearward-firing Brownings to starboard. The new nose sloped away more steeply, affording the pilot a better view forward and downwards over the nose, which added to the type's later popularity as a training aircraft.

The Blenheim V was not felt to offer sufficient advantage over the Mk IV for it to be considered suitable for operations over occupied Europe, so the last four Blenheim units, Nos 18 and 114 in No 2 Group and Nos 13 and 614 in Army Co-operation Command, re-equipped with the aircraft, but were immediately transferred to the Mediterranean to support Operation *Torch*. No 139 Sqn also briefly flew the Blenheim V, but only while transitioning to the Mosquito. Some Blenheims remained active in the UK, but these were almost all Mk Vs operating almost exclusively in the target facilities and advanced training roles.

The last home-based Blenheim units were No 17 SFTS at Cranwell (and at Spittlegate from May 1945), and No 9 (Pilots) Advanced Flying Unit at Errol and No 12 (Pilots) Advanced Flying Unit at Spittlegate. The last pilots to fly the Blenheim found it delightful. Stripped of operational equipment and armour, even the Blenheim V felt like a 'hot rod' compared to the Airspeed Oxford, and if you didn't have to risk your neck going to war in it, there is no doubt that the Blenheim made a fine aerial conveyance – a gentleman's carriage of the air, which is, after all, what it had originally been designed as. Had war not broken out, there is no doubt that the Blenheim would be fondly remembered as a peacetime light bomber in the same mould as the Hart, but unfortunately, in the harsh world of combat, it was found wanting, and many fine and brave young men met their untimely ends in it.

FIGHTER COMMAND

If the Blenheim proved inadequate as a bomber, it offered even less potential as a frontline fighter. But while it did suffer certain weaknesses, it was also 'better than nothing' (and a vast improvement on the obsolete Demon), and thus marked a useful interim aircraft prior to the introduction of the Boulton Paul Defiant. In fact, the very concept of the two-seat fighter was an anachronism, and even ACM Dowding saw little virtue in the type, although as the Munich Crisis unfolded he was only too aware that four of his precious fighter squadrons were still equipped with open-cockpit biplane Demon fighters.

The conversion of redundant Mk I bomber Blenheims into fighters offered a chance to convert the Demon squadrons to a modern monoplane fighter, and one whose top speed was famously (if inaccurately) compared to that enjoyed by enemy fighters. In November 1938, the decision was taken to re-equip Nos 23, 29, 64 and 25 squadrons (in that order) with 19 Blenheims each. It was originally planned that these would consist of 14 fully converted Blenheim IEs and five interim Blenheim IRs. Almost immediately, it was also decided that Nos 600, 601 and 604 Sqns would also re-equip with Blenheims. In addition to squadrons intended to become operational with Blenheim fighters, other units received the aircraft as interim equipment, pending receipt of Hurricanes or Spitfires.

This interim advanced training role was an important one, and many of the expansion period squadrons raised on Blenheim IFs never became operational on the aircraft, transitioning onto Spitfires or Hurricanes via the Blenheim. More importantly, especially for the handful of units destined to use the Blenheim fighter 'long term', the aircraft's range and endurance made it suitable for the air defence of coastal shipping, freeing

Blenheim IF fighters of No 604 'County of Middlesex' Sqn lined up at Northolt in January 1940. The aircraft soon had their two-letter squadron codes relocated aft of the roundel, with an individual letter ahead of it. The squadron originally applied these codes in the appropriate flight colour, but they were toned down following the outbreak of war (*via Phil Jarrett*)

With a tarpaulin cover over its dorsal turret, a Blenheim IF of No 25(F) Sqn awaits preparation for its next mission at North Weald in February 1940. Unusually, the aircraft does not appear to have a gunpack fitted, but instead sits with its bomb doors open. Blenheim fighters quickly lost their rudder serials after September 1939

up single-seat monoplane fighters for 'more important' tasks.

The fighter Blenheim was also well-suited to long range intruder missions, although the type's poor performance by comparison with single-seat fighters meant that missions usually had to be undertaken under the cover of darkness. But in the early months of the war, Blenheim intruders operated by day and by night, with a surprising degree of success.

There was initially disagreement as to exactly what the Blenheim fighter should be, with Dowding (and No 11 Group) arguing for a single-seat aircraft, with forward-firing armament only, or with the observer providing a lookout from a lightweight, Beaufighter-type, cupola, rather than from a heavy, power-operated, gun turret. A single-seat Blenheim might have enjoyed a 10-15 kt speed advantage over a turreted version.

The bomber Blenheim was of course inadequately armed for the fighter role, with only a single Browning in one wing, and with a Vickers K gun in the mid-upper turret. To these were added four more rifle-calibre machine guns in a ventral pack mounted in the former bomb bay, with 500 rounds of belted ammunition per gun. This marked a major improvement over a single 0.303-in machine gun, but it was not enough. Even with eight such guns, the early Spitfire proved lacking in punch. Against modern aircraft, which proved able to survive dozens (and even hundreds) of hits by 0.303-in rounds, a 20 mm cannon was essential. Unfortunately, no-one seemed to realise this, even though Blenheim I fighters (converted from surplus RAF Mk I bombers) exported to Yugoslavia mounted a pair of 20 mm cannon. Moreover, the Blenheim's bulky and angular gun-pack generated a significant amount of drag, knocking the aircraft's top speed down from 280 mph to about 255 mph.

Standard, unconverted Blenheim I bombers in the MUs (maintenance units) began to be allocated to fighter squadrons in December 1938, and deliveries began later the same month. Although a hastily-modified prototype fighter conversion was already flying at Martlesham Heath, a full trial conversion (of L1512) was not started until 26 January 1939. The original 150 gun packs for Fighter Command's new Blenheim fighters were manufactured by Southern Railways at their Eastleigh workshops, and were fitted to the former bombers at ASUs(aircraft storage units) and frontline stations by service personnel and contractor's working parties.

It was originally agreed that the permanent fighters (now known as Blenheim IFs rather than IEs) would have provision for a cupola to replace the turret, and that the pilot would have ring-and-bead and GM

2 gunsights. The gunsights were fitted, but in January 1939, the Assistant Chief of the Air Staff (AVM Sholto-Douglas) decreed that the turrets would be retained and not removed.

Many more Blenheim gun packs were produced under subsequent production contracts, but most of these were fitted to Blenheim IVs to produce Blenheim IVFs, mainly for Coastal Command. Since these aircraft were primarily used in the shipping protection role and in an air-to-ground, long-range fighter-bomber role, their operations are described separately. Similarly, some overseas-based bomber squadrons converted some or all of their aircraft to fighter configuration. A handful of Mk IVFs were operated by Fighter Command, most notably by Max Aitken's No 68 Sqn. But even within Fighter Command, Blenheim fighters were used as much for shipping protection, convoy escort and intruder duties as they were for air defence.

One Blenheim (L1348 of the Photographic Development Unit at Heston) was later used for extensive performance trials by the RAE with a number of modifications and armament options, including clipped wingtips, Mercury XV engines and 10° de Havilland constant speed propellors, and with turrets (extended and retracted), belly gun packs, and a twin Browning gun ring behind a small wind shield. These 'mods' did give a useful improvement in performance, but it was not sufficient to justify widespread incorporation on frontline Blenheim fighters, not least since the Beaufighter was, by then, just around the corner. At the RAE, L1290 was tested with various mid-upper turret and gun ring configurations, and this led to Fighter Command eventually deciding (on 2 October 1940) to replace the Blenheim fighters' heavy turrets with plywood fairings, with twin Brownings on a ring mounted in the forward hatch, protected by a small windshield, but by this time the Blenheim's career in Fighter Command was almost over.

Although the Blenheim IF was fully equipped for night flying, the type was not initially seen as being a dedicated nightfighter, and reflections in the heavily glazed nose made the aircraft difficult to fly nocturnally. When an aircraft was needed to train airborne radar operators from Bawdsey, however, a Blenheim was detached to Martlesham Heath, and proved sufficiently capacious to carry the bulky new AI radar equipment. The success of this first radar-equipped Blenheim led directly to the conversion of further aircraft to serve as radar-equipped nightfighters. Experimental airborne radar sets were first flown in 1937, and by the end of

This Blenheim IF served with No 600 Sqn, and is seen at Manston during the summer of 1940. The application of fin flashes was necessitated by operations over France, since all *Armée de l'Air* aircraft wore prominent rudder stripes, and coloured tail stripes were felt to be an easy and reliable recognition feature (*via Andrew Thomas*)

Although of poor quality, this view of a No 23 Sqn Blenheim IF clearly shows the half-white/half-black undersides adopted by RAF fighters as a recognition aid during late 1938 – underwing serials were removed before war broke out. The retention of a black cowling, tailplane and aileron on the starboard side was routine. Essentially, only the underside of the starboard wing itself was repainted white on most aircraft (*via Andrew Thomas*)

63

1938, contracts had been placed for the production of such equipment. It became increasingly clear that an AI-equipped fighter should be twin-engined, and have a two-man crew. In the absence of alternatives, the interim choice came down to the Blenheim IF. On 17 July 1939 a secret minute called for 21 Blenheim IFs to be equipped with radar, and equipment was rushed to the RAE from the Pye and Metrovick factories. The radar was installed at the RAE, and deliveries of AI-equipped Blenheim IFs to No 25 Sqn began on 31 July 1939. Some 15 AI-equipped aircraft were on charge by the time war broke out.

In November 1939, No 600 Sqn received three AI-equipped Blenheim IFs, and these equipped a detached flight at RAF Manston, commanded by Sqn Ldr (Later Air Marshal) Walter Pretty, who had previously commanded one of the first Chain Home Low radar stations. This unit in turn formed the basis for the Fighter Interception Unit at RAF Tangmere, which operated Blenheims fitted with a succession of generations of AI radar, and which pioneered radar nightfighting tactics and techniques, as other units began to re-equip.

But despite these developments, the Blenheim IFs of Fighter Command had remained primarily dedicated to convoy protection and long-range fighter-bomber duties. On the day that the war began, Nos 29 and 604 Sqns each despatched six Blenheim IFs over the North Sea to patrol in search of inbound German raiders, but found none. On 28 November 1939, however, Fighter Command's Blenheims finally hit the headlines when six Mk IFs from both Nos 25 and 601 Sqns flew 250 miles across the North Sea and strafed the Luftwaffe seaplane base at Borkum – none carried dinghies and no crewmen wore Mae Wests. All returned safely.

Ironically, the first radar-directed kill by a Blenheim was not made by a squadron aircraft. Instead, the aircraft assigned to calibrate the Chain Home radar station at Bawdsey (and based at Martlesham Heath), and later fitted with an experimental radar installation, was vectored on to a He 111 of KG 27 by Wg Cdr W R Farnes at Bawdsey on the morning of 5 February 1940. Flt Lt Christopher D S 'Blood Orange' Smith attacked the enemy aircraft, guided by Farnes and his own radar man, AC 1 A W Newton, but made the mistake of following it down in an effort to confirm his kill. Before the He 111 hit the sea its gunner opened fire, hitting Smith in the chest and upper arm. He managed to make a forced landing at Martlesham Heath and subsequently recovered from his injuries, while his aircraft (Blenheim IVF P4834) burned out on the airfield. But radar,

No 219 Sqn operated the Blenheim IF for one year, flying from Catterick on mainly convoy escort duties. This aircraft has the usual black and white undersides, although they are divided along the centreline, and not at the starboard wing root (*via Andrew Thomas*)

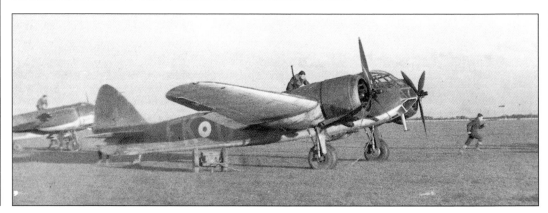

This all-black Blenheim IVF was the personal aircraft of Wg Cdr Max Aitken during his period as the commanding officer of No 68 Sqn. Radar-equipped, the aircraft had previously served with both the FIU at Ford and No 600 Sqn at Manston (*via Bruce Robertson*)

and especially AI radar, was still in its infancy, and it was to be some time before frontline radar-equipped aircraft were able to score kills.

The start of the *Blitzkrieg* in the West saw Fighter Command Blenheims committed once again. On 10 May Sqn Ldr Anderson of No 600 Sqn engaged several He 111s at 0340, returning to crash land at Manston with his aircraft's turret, flaps and undercarriage disabled.

Later that morning six No 600 Sqn Blenheim IFs were sent to attack German transport aircraft as they unloaded at Waalhaven, it being felt that bombers might have posed more danger to Dutch civilians. Five of the aircraft were shot down by Bf 110s, but one, flown by Flg Off Norman Hayes, returned to base successfully after destroying a Ju 52 on the ground and another in the air, despite having received damage from the Bf 110s. Finally, he attacked the Messerschmitts, breaking up their formation. Hayes and his gunner, Cpl G H Holmes, were respectively awarded a DFC and a DFM for their actions. Another crew survived a forced landing, whilst the gunner and observer in the lead aircraft (the only Blenheim IF carrying three crew) also escaped with their lives.

Less than two hours later, No XV Sqn, with its Blenheim bombers, also attacked Waalhaven, causing further damage. After lunch No 600 Sqn sent out four more of its Blenheims, damaging a He 111, before returning to base, relieved on patrol by No 25 Sqn. No 604 Sqn was also operational on the 10th, escorting Wattisham Wing Blenheims as they bombed Ju 52s unloading at Waalhaven and on beaches north of the Hague. No 604 claimed four Ju 52s destroyed and three damaged, but lost an aircraft – the crew 'walked home' after setting fire to its Blenheim.

On 12 May, three Blenheim fighters of No 235 Sqn were engaged by Bf 109s and Bf 110s while covering the evacuation of the Dutch queen by Royal Navy destroyers and British Royal Marines. Two of the Blenheims were shot down, but they themselves downed a Bf 109 and a Bf 110. Although the Blenheim fighter was out-gunned and out-performed by the Bf 109, there were areas in which the former enjoyed some advantages, and when properly appreciated and exploited, an experienced crew could sometimes turn the tables on their attackers. Plt Off Hugh Wakefield, an observer with No 235 Sqn pointed out that 'a Blenheim could turn inside a '109, if we knew precisely when to begin the turn'.

On 13 May, the Blenheims of Nos 235, 248 and 254 Sqns were in action, still as Fighter Command units, although they were soon to trans-

fer to Coastal Command. No 600 Sqn returned to Hendon on 15 May, after suffering heavy losses, and was replaced at Manston by No 604 Sqn. It soon became clear that the Blenheim IF was too vulnerable for ground strafing by day, and the squadrons began switching to night intruder operations and nightfighter patrols. Before France fell, Blenheim nightfighter operations accounted for a He 111 claimed as a probable by Flg Off Hunter of No 604 Sqn on 18 May.

When the Luftwaffe switched to night attacks during the Battle of Britain, the Blenheim fighter was one of the only counters available to Fighter Command. While radar remained just over the horizon, the Blenheims began their night-time vigil, relying on eyesight and searchlights, with generally frustrating results, although there were exceptions. On the night of 18/19 June, for example, seven No 23 Sqn Blenheims (and one from No 29) found aircraft from a 70-strong raid by He 111s. As soon as they fired, the Blenheims gave away their own position, making them vulnerable to heavy counter-fire, but three aircraft nevertheless scored victories. Unfortunately one of these aircraft was forced to ditch after being hit by fire from its victim, the No 29 Sqn Blenheim pilot being drowned – another Blenheim was also lost to counter-fire.

But while the end was in sight for the Blenheim nightfighter, there were still laurels to be added to its crown, especially by the radar-equipped aircraft. On 19 July No 600 Sqn pilot Flt Lt David Clackson actually found an enemy aircraft as a 'blip' on his set, but was unable to engage. No 600 Sqn's Plt Off 'Archie' McNeill Boyd engaged He 59s on 20 July and 25 July, albeit without success. Finally, on the night of 22/23 July 1940, the FIU (Fighter Interception Unit) scored its first kill using radar, the honour falling to Flg Off G Ashfield, his gunner, Plt Off G Morris and radar operator, Sgt R H Leyland, flying a Blenheim fitted with the latest AI Mk IV. Their victim was an inbound Do 17Z of 2./KG 3, which fell into the

This well-worn Blenheim IVF spent time with No 143 Sqn, which served as a nightfighter training unit in Northern Ireland before re-equipping with Bristol Beaufighters in September 1942 (*via Peter H T Green*)

channel. This marked the beginning of a brief spell of success for radar-equipped Blenheims, although the type continued to operate by day as well. In the earliest days, the radar operators were regarded very much as supernumeries on fighter squadrons where two-man crews were the norm. The men were usually aircraftmen or LACs, and were only later made up to sergeant, like the majority of the fighter W/Op Air Gunners.

On 11 August, two Blenheims of No 604 Sqn (escorted by three No 152 Sqn Spitfires) attacked ships and an He 59 on the sea 30 miles off the French coast – they all fought their way back despite the intervention of six Bf 109s. On 11 September the squadron flew a similar sortie, attacking a Do 18 under tow by German E-Boats.

In the wake of a daylight raid against Driffield and Scarborough on 15 August, 12 of No 219 Sqn's Blenheims engaged the enemy bombers without success. One pilot, Sgt Dube, was wounded, but was assisted in a wheels up landing at Driffield by his gunner, Sgt Bannister. Both men won DFMs in the process.

By August, Manston was home to the FIU and No 600 Sqn, and the airfield was heavily bombed and strafed on the 12th, 14th, 16th and 20th of the month. The last raid (by ten Bf 109s) was opposed by three No 600 Sqn Blenheim IFs, although their efforts were frustrated by wild firing from the airfield's Bofors anti-aircraft guns.

KG 53 lost a He 111 to Plt Off R A Rhodes and Sgt Gregory of No 29 Sqn on the night of 17/18 August. They picked up the intruder near Chester and followed it to Spurn Head, where they engaged the bomber and sent it spinning into the sea off Cromer. On 20/21 August Flt Lt J Adam and Plt Off Watson of No 29 downed another raider off the Isle of Wight, catching the aircraft (probably a Ju 88) only when it circled to get its bearings. Three days later No 29 Sqn's Flt Lt 'Bob' Braham (subsequently to become the RAF's leading nightfighter ace) found a He 111P (from III./KG 55) in Hull's searchlights whilst flying Blenheim IF L1463, and the two aircraft exchanged fire. Searchlight crews subsequently reported seeing the German aircraft burning on the sea.

On 4 September Plt Off Rofe of No 25 Sqn attacked three night raiders, but his aircraft was damaged by friendly anti-aircraft fire before he could conclude the engagement successfully. His squadron-mate, Plt Off M Herrick had more success early the next morning, despatching a 1./KG 1 He 111 near Bentwaters and claiming another as a possible. On 14 September Herrick downed another He 111 near Sheering, making him the most successful Blenheim fighter pilot with two kills and a possible! No 600 Sqn downed a Ju 88 (claiming it as a 'Heinkel') off Bexhill on 15/16 September, which was credited to their most successful Blenheim pilot, Flt Lt C Pritchard.

At the end of the Battle of Britain the RAF had six Blenheim units (Nos 23, 25, 29, 219, 600, and

The final colour scheme for Blenheim nightfighters was overall black, as seen on this No 54 OTU Blenheim IF. The aircraft is fitted with AI Mk III radar, whose antennas can be seen on the nose and above and below the port outer wing (*via Phil Jarrett*)

604), two Defiant squadrons and one Hurricane unit with which to counter the Luftwaffe's night blitz. Expansion of the nightfighter force was completed using Hurricanes as interim equipment, but the Blenheims (relatively few of which had AI radar) were earmarked for replacement by fully radar-equipped Beaufighters, which had entered service with the FIU in August 1940, and with Nos 23, 25,

29, 600 and 604 Sqns in September. The Beaufighters trickled to the squadrons in ones and twos, and operated alongside Blenheims for some time, with Nos 25 and 604 becoming the first fully-equipped Beaufighter units when they finally withdrew their venerable Blenheims from operations in January 1941.

The first Beaufighter had actually been delivered to an MU for the fitting of operational equipment on 27 July 1940, and the new type had begun operations with the FIU on 12 August 1940 and with No 604 Sqn on 2 September. Although the Blenheims lingered on until May 1941, they became increasingly sidelined.

The RAE trials referred to earlier indicated the advantages to be obtained by removing mid-upper turrets. On 2 October 1940, Fighter Command ordered that the 64 aircraft equipping its four Blenheim IF squadrons should be thus modified, purchasing 100 modification kits. No 604 Sqn began modifying its aircraft that month, with Nos 25, 219 and 23 following suit, albeit slowly. There was little urgency, not least because the Beaufighter was by then on the verge of entering service, offering heavier armament and higher performance.

When the Luftwaffe attacked Coventry on the night of 14/15 November, Blenheims bore the brunt of the defensive effort, flying 35 sorties, although Beaufighters flew 12, Defiants 30 and Hurricanes 43. John Cunningham's famous victory on 19 November was scored in a 'Beau, marking the new type's first kill and drawing a line under the Blenheim's tally. Thereafter, the Beaufighter began to replace Blenheims at an accelerating rate.

From 21 December, Blenheim nightfighters began flying intruder missions against Luftwaffe bomber bases, alongside other nightfighters. Blenheims began Operation *Intruder* when six Blenheim IFs of No 23 Sqn attacked bomber airfields in Normandy.

While the long-serving Blenheim nightfighter units busily converted to the new Beaufighter, 7 January 1941 actually saw the formation of a new Blenheim nightfighter squadron, No 68, at Catterick, under the leadership of Hurricane ace Wg Cdr the Honourable Max Aitken (see *Osprey Aircraft of the Aces 18 - Hurricane Aces* for more details). The squadron moved to High Ercall in April 1941.

If the Blenheim fighter achieved little in concrete terms, it did at least pioneer the tactics and equipment later used to great effect by nightfighting and night-intruding Bostons, Beaufighters and Mosquitos, and proved a thorn in the side for German night bomber aircrew.

The wartime censor obliterated the underwing radar antenna on this photo of K7159 (see on the previous page), but left the overwing and nose-mounted antennae. The individual aircraft letter 'N' was outlined in yellow, denoting the aircraft's assignment to an OTU (*Aerospace Publishing/Wings of Fame*)

COASTAL COMMAND

W hen war broke out Coastal Command was poorly equipped to meet its responsibilities, with flying boats of various types for long-range patrol, ASV and ASW duties, and with Ansons and a handful of Hudsons for inshore and convoy escort duties. During the 'Phoney War' these were adequate, since brushes with the enemy were fortunately rare, but once the *Blitzkrieg* began, the Luftwaffe began to systematically interdict shipping from the Shetlands to the Thames Estuary.

Coastal Command clearly needed new, more capable, convoy escort and strike fighters, not least to equip the four 'trade protection' squadrons first envisaged during the earliest days of the war. Every single-engined fighter was needed by Fighter Command (even the useless Defiants), and aircraft like the Spitfire and Hurricane in any case lacked the endurance required to protect shipping sailing between the Shetlands and the Thames Estuary, even in coastal waters. Before the invasion of France this was felt to be a virtually risk-free task, with little danger of interference by single-engined fighters, and aircraft like the Blenheim were felt to be capable of dealing with multi-engined types like the He 111 and Ju 88.

Coastal Command evaluated a Blenheim IV (L4846) with No 217 Sqn between March and August 1939, and gained its first operational Blenheims in October 1939, when No 233 Sqn at Leuchars stood up 'D Flight' at Bircham Newton. This unit is often overlooked since it was never fully equipped with Blenheims, instead being absorbed by one of the former Fighter Command Blenheim fighter units in January 1940 when these transferred to Coastal Command. As so often, the Blenheim proved a useful interim answer to a difficult problem, and four Trade Protection units were formed within Fighter Command during late 1939. These squadrons transferred to Coastal Command in February 1940, fulfilling much the same task, but now operating from Coastal Command aerodromes and officially designated as 'fighter reconnaissance' units.

Initially equipped with a mix of Blenheim Is and IFs, from April 1940

A busy seen at RAF North Coates in March 1940, with the Blenheim IVFs of Nos 235 and 254 Sqns lined up ready for flight. The two squadrons had transferred to Coastal Command in February and January 1940, respectively, from Fighter Command. Blenheim IVFs used by Fighter Command had retained day bomber colours (without black/white undersides) and these colours were originally also used by the Coastal Command units (*via Andrew Thomas*)

the four squadrons rapidly standardised on the long-nosed IVF, which operated with a third crew member. Whereas the non-radar versions of the IF had a crew of only two (pilot and gunner) the IVF had a pilot, wireless operator/air gunner and an observer. No 248 Sqn had both versions on charge when it transferred from Hendon to North Coates and to Coastal Command control, but immediately discarded the IFs. The latter type lasted until March with No 254, which transferred from Stradishall to Bircham Newton, where it absorbed No 233 Sqn's 'D Flight'. No 235 Sqn relinquished its IFs in May 1940, after transferring from Manston to North Coates. No 236 Sqn moved from Martlesham Heath to North Coates, and similarly used both Blenheim variants, retaining the Mk IF until August – the unit kept its short-nose Blenheims longer than the other squadrons because it retained a secondary day fighter role.

The new units were soon on the move, and in April No 235 Sqn moved to Bircham Newton, No 236 to Speke, and No 248 to Thorney Island. No 254 remained at Bircham Newton but soon established detachments at Detling and Thorney Island. Detachments were also introduced at Lossiemouth, to allow operations along the Norwegian coast.

The line dividing Coastal Command Blenheim fighter operations and those carried out by Fighter Command Blenheims was sometimes a fine one, since aircraft from both Commands sometimes carried out the same type of operation. Often, the only difference was in the command and control arrangements, and interestingly No 248 Sqn was actually transferred back to Fighter Command on 22 May 1940, joining No 13 Group for Scottish coastal defence work, operating from Montrose and Dyce – it rejoined Coastal Command on 26 June, remaining in Scotland. On 25 May No 236 Sqn also transferred back to Fighter Command and moved to Filton, returning to Coastal Command at Thorney Island in July.

Within Coastal Command the Blenheims carried out a range of missions, these depending on where a unit was based. Perhaps the key early responsibility was protecting the North Sea fishing fleet. Although the four Blenheim squadrons all began operations from English airfields, two units (Nos 248 and 254) soon transferred to Scotland and began flying defensive missions off the Scottish coast, and then quickly began mounting photographic and reconnaissance missions, and later offensive sweeps, to the Norwegian coast, especially around Trondheim.

Even before it moved to Scotland, No 254 Sqn had experienced action. While flying a fishery protection sortie on 9 March two crews engaged a He 111, although unfortunately the enemy aircraft fractured the spar of L8766, hit L8841 and then escaped. Another squadron aircraft was damaged the next day, this time by fire from the ships it was escorting!

During April, No 254 Sqn's Blenheims provided detachments at Lossiemouth, and from here they flew many missions in support of the allied forces involved in the Norwegian campaign, including an airfield attack against Stavanger on 10 April, and another against Vaernes. The Stavanger attack (by three Blenheims) resulted in a claim of three enemy aircraft destroyed, one of them a Ju 88. On 11 April the unit successfully strafed a German destroyer in Hjelte Fjord, whilst a fortnight later Plt Off Illingworth downed a He 111 during a sortie flown from the squadron's new home at RNAS Hatston, on the Orkneys.

The two squadrons which remained in England took rather longer to become operational within Coastal Command, No 235 Sqn finally flying its first sorties in May 1940 and No 236 in June. No 254 Sqn, on the other hand, had begun offensive sweeps in March 1940, while No 248 had been declared operational in June.

On 2 May No 235 Sqn flew Coastal Command's first sortie against shipping off the Dutch coast, and three days later it flew a reconnaissance to Wilhelmshaven, where two of the unit's Blenheims had a narrow escape from three Bf 109s – one aircraft returned home with an engine put out of action. On the 7 May three squadron aircraft covered a mining operation by Beauforts and Swordfish off Nordeney. Such sorties were soon to become far more dangerous when Hitler invaded the Low Countries as a precursor to his assault on France.

The campaign against the Germans in Norway continued even as Wehrmacht tanks warmed their engines prior to the *Blitzkrieg* in the West. On 9 May No 254 Sqn escorted Fleet Air Arm (FAA) Skuas when they bombed Bergen, sinking two enemy ships. Three days later two squadron aircraft beat off about 17 enemy bombers as they tried to interfere with British minelayers operating off Flushing. From 13 May, the unit began covering the evacuation of the Hague, and was committed to operations off the Dutch coast. Flt Lt Pennington-Leigh put his four-gun gunpack to good use on 18 May by shooting down a pair of Bf 110s.

As it became clear that the Battle of France was nearing its end, the threat of an invasion of Britain became more real. Accordingly, two Coastal Command Blenheim units were transferred to Fighter Command for anti-invasion patrols. No 248 Sqn moved from Thorney Island to Montrose and Dyce on 22 May, while No 236 left Speke for Filton.

Activity off Norway remained intense, even though attention had naturally switched to the south. No 235 Sqn (by now at Detling) was heavily involved in the Dunkirk evacuation, moving to Thorney Island to cover the retreat from the more westerly French ports. Meanwhile, off Norway 20 June was a good day for No 254 Sqn, as three of the unit's aircraft attacked a U-Boat and shot down a He 115.

The two squadrons lost to Fighter Command returned to the fold on 26 June and 4 July. This was perhaps just as well, since attrition was mounting steadily. Six aircraft from No 235 Sqn were bounced by at least a *Staffel* of Bf 109s on 27 June while carrying out a SAR operation, and only two returned to base. On 6 July the vulnerability of the Blenheim in Norwegian airspace was also demonstrated (to those who hadn't noticed what was happening over France and in the channel) when two fell to Bf 110s as they escorted HMS *Coventry* and HMS *Southampton* 50 miles off Stavanger.

Following the withdrawal of British forces from France, Coastal Command gained two new Blenheim-equipped units in July 1940, Nos 53 and 59 Sqns, both of whom were former army co-operation and reconnaissance-bomber units attached to the Air Compo-

Coastal Command used the Blenheim IVF (with its underfuselage gunpack) in preference to the 'standard' Blenheim IV day bomber. The aircraft proved no more effective than No 2 Group Blenheims used in much the same role, however, although they did enjoy some success against enemy reconnaissance aircraft. Although the gun pack of the Blenheim fighter was installed over and within the former bomb bay, this did not prevent the Mk IVF from carrying bombs, which were instead mounted on external bomb carriers below the fuselage, as seen on this No 235 Sqn aircraft, taking off from RAF Bircham Newton (*via Peter H T Green*)

RAF Manston was a busy airfield in 1940, housing day and night fighters involved in the Battle of Britain, and bombers engaged in hitting enemy invasion barges and shipping. These Blenheim IVFs wear the markings of No 235 Sqn (*via Phil Jarrett*)

Pock-marked by bullet holes, this aircraft served with No 59 Sqn until December 1940, when it hit trees while overshooting at Thorney Island. No 59 Sqn transferred to Coastal Command in July 1940 – this photo was taken a month later (*via Andrew Thomas*)

nent of the BEF. Both squadrons flew standard bomber Blenheim IVs from Detling, Manston, Thorney Island and Bircham Newton, and retained these after transfer to Coastal Command whilst continuing their primary role of strategic reconnaissance, but soon also mounting attacks on enemy ports, shipping and coastal targets.

No 53 Sqn, initially based at Detling as part of No 16 Group, covered the eastern end of the channel, while No 59, at Thorney Island (No 15 Group), watched over the westernmost part of the channel. The geographic boundary was not strictly observed, especially when bombing missions were planned, since more targets lay in the No 16 Group area. Most of the squadrons' reconnaissance missions were flown by single aircraft, which were extremely vulnerable to hostile fighters, and losses were heavy. Many recce aircraft carried bombs for use against targets of opportunity, and would attack airfields when no surface vessels could be found.

Bombing attacks by the new squadrons began on 6 July when five aircraft from No 53 Sqn attacked barges on the Ijmuiden-Amsterdam canal. The size of the Blenheim bomber raids increased steadily – on 14 July eight No 59 Sqn aircraft attacked oil storage tanks at Ghent, whilst on 1 August 13 aircraft were despatched on a medium level bombing attack against Cherbourg. The latter raid cost the life of No 59 Sqn's leader, Wg Cdr Weldon-Smith.

The only real answer to the Blenheim's vulnerability to enemy fighters was to withdraw the type from service, but this was simply not an option. Instead, efforts were made to improve the type's armament in order to make it a more costly target for the enemy to risk attacking. Accordingly, the programme to add a second machine gun in the mid-upper turret was accelerated, while pilot's reflector gunsights and swivel-mounted machine guns for the observers were also added. This, however, applied only to the Blenheim IVs and IVFs. No 236 Sqn continued to operate IFs, with two-man crews, although by mid July the squadron had equipped one of its flights with Blenheim IVs.

One of the roles assigned to No 236 Sqn was to escort civilian Short Empire flying boats flying between Poole and the Scillies, and the squadron also protected reconnaissance aircraft and flew offensive sweeps. During July the squadron had several engagements with Ju88s, but these were able to use their superior speed to make their escape, and No 236 was only able to claim damaged enemy aircraft. On 17 July, three of the squadron's Blenheims attacked three Ju 87s, but even these slow dive-bombers were able to flee. Things were even more difficult when single-engined fighters were engaged. One squadron aircraft was downed by Bf 109s on 20 July, and an attempt to exact revenge 24 hours later proved fruitless when a Bf 109 simply ran away from three Blenheims.

On 1 August Coastal Command mounted one of its biggest Blenheim operations, when 13 aircraft from No 59 Sqn bombed Querqueville airfield, supported by ten aircraft from No 236. Four of these provided top cover, and six strafed the airfield. The first wave of three strafing aircraft were all hit, two falling to accurate groundfire. Channel convoy patrols were considerably more hazardous than the sorties flown by the Scottish-based units, not least due to the proximity of fighter airfields in France.

No 236 Sqn moved further from potential trouble in August 1940, re-locating to St Eval on the north Cornish coast, and detaching its 'A' Flight to Aldergrove in September 1940. No 236 Sqn's 'B' Flight had started to move to St Eval on 2 August, almost losing an aircraft to over-enthusiastic Gladiator pilots from No 247 Sqn, who mistook the Blenheim for a Ju 88, forcing it to flee! The whole unit had completed the move to St Eval by 8 August, having shot down a Bf 109 during the intervening period.

From August, No 254 Sqn took something of a rest, flying coastal patrols and convoy escorts while No 248 took over the offensive. The increased firepower of the Coastal Command Blenheims was put to good use during August. On the 3rd of the month No 248 Sqn's Plt Off Gane attacked a U-Boat, while three aircraft from No 235 forced down an He 115. Four days later three more of No 235's Blenheims attacked and strafed a group of ships in the first deliberate, pre-planned, anti-shipping strike flown by Coastal Command's fighter Blenheims, which had hitherto flown in the convoy escort, reconnaissance, SAR and patrol roles, attacking ships only on an opportunistic basis. On 8 August three Blenheims were able to beat off attacks by 15 Bf 110s (downing one of the enemy aircraft) in a 'fighting withdrawal'), while on 11 August, another trio of Blenheims shot down a Bf 109, and drove away the enemy's wingman.

Nos 53 and 59 Sqns were extremely active during August too, attacking invasion barges, ships at

Home-based Coastal Command squadrons actually retained the Blenheim IV for longer than the Beaufort – which was effectively a Blenheim optimised to meet the Command's particular requirements! These aircraft served with No 254 Sqn (*via Aeroplane*)

Blenheim IVFs, again from of No 254 Sqn, put up a neat echelon starboard for the camera. The aircraft wear standard day bomber colours, although some Coastal Command Blenheims did eventually receive extra dark sea grey and dark slate grey upper surfaces (*Aerospace Publishing/Wings of Fame*)

sea, convoys and especially enemy airfields whenever the opportunity presented itself. Sometimes these attacks were undertaken by individual aircraft (using cloud cover to sneak in undetected) and sometimes by large, escorted formations of aircraft. The Luftwaffe responded in kind, and Detling was attacked on a number of occasions.

During September No 248 Sqn mounted a number of attacks on shipping off the Norwegian coast, bombing two vessels, and also flew anti-invasion patrols around the Scottish coast. The unit had lost an aircraft during October along with its crew, whilst the previous month had seen a Blenheim ditch after running out of fuel – the aircraft was salvaged, and the crew recovered.

September was a busy month for Nos 235 and 236 Sqns, with convoy escorts, anti-invasion patrols and escort missions for Coastal Command's Blenheim bombers, FAA Albacores, and other attack aircraft raiding French ports and coastal targets. On 11 September the two squadrons escorted an FAA raid on Calais, No 53 Sqn attacked a convoy and No 59 Sqn bombed Boulogne. All three missions were opposed by enemy fighters, but remarkably, only four of the fighter Blenheims were lost, and the remainder accounted for three Bf 109s. The three months of July, August and September had been particularly hard for Nos 53 and 59 Sqns, which lost 19 aircraft to enemy action, 13 more written off due to battle damage and two in operational accidents. This represented almost 40 per cent of total strength.

Coastal Command gained a seventh and an eighth Blenheim squadron in November 1940 when No 272 Sqn formed at Aldergrove from detached flights of Nos 235 and 236 Sqns, and No 252 Sqn was established at Bircham Newton. The latter unit was intended to be Coastal Command's first Beaufighter squadron, but operated Blenheims as interim equipment until April 1941. The ninth Coastal Command Blenheim unit arrived in December 1940 with the formation of No 86 Sqn at Gosport for convoy escort and air sea rescue duties.

The final Coastal Command Blenheim squadrons were Nos 608, 500, 404 (RCAF) and 143, which formed between February and June 1941, respectively at Thornaby, Detling, Thorney Island and Aldergrove. The two Auxiliary Air Force units were destined to fly Blenheims for less than a year, but the other two squadrons would eventually become the Command's last Blenheim operators. The formation of new Coastal Command squadrons took the force to 13 Blenheim units, but interestingly it was Bomber Command's No 2 Group which bore the brunt of the near-suicidal 'Roadstead' and 'Channel Stop' anti-shipping operations.

The rundown of Coastal Command's Blenheim force was leisurely, and began in April 1941 when Nos 252 and 272 Sqns re-equipped with Beaufighters – the latter unit moved from Aldergrove to Chivenor (with a brief soujourn at Sumburgh) for Beaufighter conversion. As noted earlier, No 236 Sqn moved west to St Eval in August 1940, but No 235 Sqn remained in the thick of it at Bircham Newton until June 1941, when it was posted north to Dyce.

By this time daylight missions over the channel and southern parts of the North Sea were simply too hazardous for the Blenheim, and Beaufighters began to take the strain. Although No 235 was replaced at Bircham Newton by another Blenheim unit, No 248 Sqn, the latter squadron was simply working up for conversion to the Beaufighter – it began flying operations off the Dutch Coast, but converted to the Beaufighter in July. One of the highlights of No 248 Sqn's final Blenheim operations came in May 1941 when a squadron Mk IV downed a Ju 88. The unit was eventually replaced at Dyce by No 235 Sqn, but the new resident unit's Blenheim IVFs flew mainly as night intruders.

Nos 53 and 608 Sqns received Hudsons in July 1941, with No 86 Sqn swapping Blenheims for Beauforts in that same month and No 248 re-equipping with Beaufighters. The latter unit was the first of the four Coastal Command fighter squadrons to trade in its Blenheims for Beaufighters, which replaced most of the Command's remaining Blenheims over the next 15 months. No 59 Sqn re-equipped with Hudsons in September 1941, however, and was followed by No 500 in November, leaving five remaining Blenheim units. No 235 Sqn gave up its Blenheims in December 1941, No 236 relinquished its last aircraft in February 1942 and No 254 in June.

No 254 Sqn had spent the second half of 1941 at Aldergrove, in Northern Ireland, flying convoy escorts in much safer skies over the eastern Atlantic, where short-range enemy fighters were less likely to be encountered. The squadron returned to Dyce for the last six months of its existence, but flew less risky missions than had been common during early 1941.

The Blenheim just outlived its stablemate, the Beaufort, in Coastal Command service at home, with the last frontline UK-based Beauforts vanishing from service in August 1942. The following month the last two Blenheim squadrons in the Command (Nos 143 and 404) finally gained Beaufighters, and the Blenheim's frontline career with the Command was over. The aircraft remained active in the convoy patrol and coastal defence roles overseas, where enemy air opposition was less fierce, but these operations are outside the scope of this chapter.

Anonymous Coastal Command aircrew struggle out to their awaiting Blenheims encumbered with flight gear in late 1940 (*via Phil Jarrett*)

MIDDLE EAST AND MEDITERRANEAN

As an interim replacement for the generation of obsolete bombers and army co-operation aircraft then equipping the RAF's front-line squadrons, Blenheim Is began to re-equip overseas units (where tactical bombers were still perceived to be useful) after only four home-based squadrons had converted to the aircraft. Outside home-based squadrons of Bomber Command's No 2 Group, nowhere was the need for modern light bombers more acute than in the RAF's Middle East Command.

The first Blenheim squadron to form in the Middle East was No 30, which replaced its Hawker Hardies at Habbaniyah from 13 January 1938 (the station was known as Dhibban until late March). By the time Mussolini's Italy entered the war on 10 June 1940, there were three more Blenheim squadrons in Aden (Nos 8, 11 and 39) and five in Egypt (Nos 30, 45, 55, 113 and 211), while No 84 had replaced No 30 in Iraq. All flew standard Blenheim I bombers, although No 113 was re-equipping with Blenheim IVs and No 30 issuing two of its flights with locally-modified Blenheim IF fighters.

Mussolini's declaration of war on 10 June was punished the next day when Nos 45, 55 and 113 Sqns sent a force of 26 Blenheims to bomb the Italian airfield at El Adem, in Libya. Two Blenheims were lost to flak and fighters, and another forced landed upon its return to base. The RAF continued to mount offensive sorties, but was unable to replace aircraft losses due to the demands of Blenheim units engaged in the Battle of France.

When Italy invaded Greece through Albania, the Greeks refused the offer of British military assistance, with the exception of air forces. Accordingly, No 30 Sqn was detached to Eleusis to help defend Athens on 1 November 1940, where it was joined by one flight from No 84 Sqn,

As Blenheim IVs replaced Blenheim Is with home-based bomber squadrons, the redundant Mk Is were transferred either to Fighter Command or to overseas units. Here, L6655 is seen arriving at Khormaksar, Aden, for use by No 8 Sqn in the spring of 1939. The observer can be seen leaning out of the top hatch as the aircraft taxies in (*via Bruce Robertson*)

Blenheim Is of No 45 Sqn are seen lined up at at Mersah Matruh, freshly re-painted in desert camouflage (with tan replacing the normal dark green) ready for an inspection by King Farouk, and prior to a detachment to the Sudan
(*via Andrew Thomas*)

which moved to Menidi (formerly Tatoi) with the rest of the unit, and alongside No 211 Sqn. No 11 Sqn moved to Larissa in mid January 1941, and No 113 Sqn followed (to Menidi) in March. The British forces in North Africa were able to spare the Blenheim units largely due to the success of General Wavell's offensive against the Italians.

While the Italians were hardly high quality opposition, their failure prompted German intervention in both the Balkans and North Africa. When Germany invaded Greece on 6 April, British forces were rushed in (mainly from North Africa) to try and stem the tide. Unfortunately the campaign went badly, and RAF units committed suffered heavy losses.

No 113 Sqn lost all of its surviving aircraft during a German attack on its airfield on 15 April 1941, withdrawing to Palestine to re-equip. Things went no better for the other RAF units committed to this unequal struggle. On Easter Sunday (13 April) one of No 211 Sqn's Blenheim Is was sent on a dangerous reconnaissance mission to the Albanian ports of Valona and Durazzo, while the remaining seven serviceable aircraft mounted an attack on German MT at Prilep, on the Bulgarian border.

The pilot of the recce aircraft, Jack Hooper, encountered no opposition, but the other six aircraft were bounced by Bf 109s just short of their targets and all were shot down. Two crewmen bailed out, and buried those of their comrades they could find before struggling to Larissa, where they hitched lifts in a pair of Lysanders. These too were shot down on take-off by Bf 109s, leaving just one survivor from No 211 Sqn's abortive ground attack mission, Flt Lt Alan Godfrey. The remaining aircraft from most of the Blenheim squadrons were withdrawn to Egypt at the end of

Several of No 45 Sqn's Blenheim Is were named after British pubs. This aircraft, flown by Flg Off Rixon, and seen over Agordat in Italian East Africa in July 1940, was *The Cheshire Cheese*
(*via Andrew Thomas*)

April, with the newly arrived No 203 Sqn withdrawing in May. No 30 Sqn had been posted to Crete to protect shipping involved in the evacuation, and this unit also withdrew to Egypt at the same time.

No 30 Sqn got away from Crete just in time to avoid Operation *Merkur* – Germany's invasion of the island by elite paratroops and glider- and Ju-52-borne infantry and mountain troops. Egyptian-based Blenheims did fly sorties

No 30 Sqn received Blenheim I bombers at Dhibban (renamed Habbaniyah on 25 March 1938) as early as January 1938, before moving to Ismailia. The squadron then converted the aircraft within 'B' and 'C' Flights to fighter standards in June 1940. In Greece, the unit flew mainly in the bomber role, but re-converted its aircraft to fighter standards when it returned to Egypt (*via Bruce Robertson*)

This Blenheim I of No 211 Sqn is seen landing at Tatoi after a raid on Italian positions in Albania during late 1940. The aircraft wears typical Desert Air Force colours, and is fitted with Vokes filters below the engine cowlings. RAF Blenheims deployed to Greece were decimated when German forces invaded (*via Peter H T Green*)

against the invading German forces, but suffered heavy losses and failed to prevent the eventual fall of the island and its garrison.

The early British successes against the Italians in North Africa sowed the seeds for a more difficult campaign. The inability of the Italians to combat British forces led to the deployment of the Afrika Korps, which was supported by its own air army, equipped with modern fighters. The deployment of German forces in North Africa rapidly halted and then reversed General Wavell's offensive, and the British began retreating back towards Egypt again.

Part of the reason for the retreat from Libya was the fact that British forces had briefly been forced to fight on two fronts. In April 1941 dissident Iraqi generals, funded and prompted by Germany, staged a coup against the King and his government. The King was evacuated by the RAF to its main base in Iraq, Habbaniyah, while British forces were rushed into Basra and Shaibah. There was an immediate need to counter the threat to the strategic route to India, and Iraq's vital oil producing facilities, which appeared to be in imminent danger of falling into pro-German hands.

Habbaniyah was besieged by Iraqi ground forces, and came under air attack from German aircraft (mostly operating in Iraqi markings). The massive aerodrome was the home of No 4 Flying Training School,

Blenheim IV N3589, formerly of No 40 Sqn, is seen in Italian markings after landing in error on Pantellaria Island on 13 September 1940 whilst en route to Egypt (*via Phil Jarrett*)

equipped with obsolete types used for training, but these were hastily pressed into service to support the troops holding Habbaniyah's perimeter, and were joined by Wellingtons and the Blenheim IVs of No 203 Sqn. At the aerodrome at H4 in Jordan (a pumping station on the main H oil pipeline) flights from the Blenheim-equipped Nos 84 and 203 Sqns began operations aimed at denying the use of Iraqi aerodromes at Ar Rutbah and H3 to the Luftwaffe.

At the same time that Iraqi generals mounted their coup, further trouble erupted in Syria, a French colony under Vichy control. Syria represented a useful potential base for the Axis, lying in a position which threatened Iran and Iraq, and the route to India. The Vichy regime eagerly granted refuelling facilities to Luftwaffe aircraft involved in operations against the British in Iraq, and also allowed Syrian airfields to be used to bring in arms and supplies to the Iraqi rebels. The Vichy French reinforced Syria's air defences, sending a squadron of Dewoitine D.520s from Algeria to Rayak, hoping that this would deter British air attacks. In fact, aircraft from H4 immediately began attacking German transports as they unloaded at Syrian aerodromes, beginning on 14 May when Flg Off Watson of No 203 Sqn spotted Ju 90s and other aircraft at Palmyra during a long range recce mission. Later that day, Watson, accompanied by two No 84 Sqn Blenheims and a gaggle of Curtiss Tomahawks, returned and strafed the airfield.

Meanwhile, British air operations in Iraq eventually resulted in the defeat of the rebel Iraqi forces, and allowed British and Indian troops to advance on Baghdad. The new ruler of Iraq fled the country and the King was reinstalled on 1 June.

By the end of May it was becoming clear that the British were about to defeat the Iraqi rebels, and in the hope of forestalling an allied invasion, the Vichy High Commissioner of Syria requested that Germany should withdraw all forces from the country. The Germans acceded and left by 8 June, but it was too late for on that day an invasion was launched from Trans-Jordan and Palestine. Blenheims conducted most of the bombing and reconnaissance sorties in support of the allied invasion, with aircraft being drawn from Nos 11, 45 and 84 Sqns. At least eight Blenheims were shot down during the campaign, three falling to D.520 fighters on 10 July during an attack on an ammunition dump at Hama. After heavy fighting,

This wrecked Blenheim IV was photographed at Heraklion, on the island of Crete, shortly before the German invasion. The bent back propellor blades indicate that the aircraft had been involved in either a landing accident in which the landing gear collapsed, or a forced landing effected with the gear in the raised position (*via Andrew Thomas*)

This precariously placed Blenheim IV of No 113 Sqn crashed at Niamata, Greece, in April 1941 after returning from a sortie with battle damage (*via Phil Jarrett*)

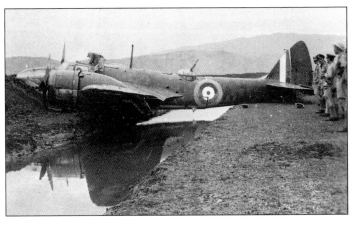

the Vichy forces capitulated on 14 July.

With Syria and Iraq safely in allied hands, the British could again turn their full attention westwards, towards the German and Italian forces in Libya.

An integral part of the campaign for North Africa was waged from, and over, the island of Malta, Britain's tiny outpost in the mid-Mediterranean proving ideally situated as a base from which to launch attacks against Axis shipping ferrying men and material from Italy to North Africa. Unfortunately, Malta was within flying range of bases in Italy, so the Luftwaffe and the Italian air force made every effort to knock out the island's aerodromes, and thus disrupt air operations being carried out from them. A more general bombardment of the island continued throughout much of the war, intended to soften up Britain's island fortress prior to invasion. As a result, all flying operations from Malta were subject to relatively heavy losses.

The deployment of Blenheims to Malta began on 26 April 1941 with a detachment by six crews from No 21 Sqn, which served as a proving flight. No 2 Group then began a regular rotation of certain of its Blenheim units for between five and six weeks each. The first of these deployments was made by No 82 Sqn in June 1941. The Malta Blenheims attacked enemy shipping, and also ranged against targets on the North African mainland, close to the coast. No 82 Sqn was followed by No 110 Sqn in July 1941, and then (later that same month) by No 105 Sqn led by Hughie Edwards. No 107 Sqn followed in September 1941, augmenting rather than replacing No 105.

Losses among the detached Blenheim squadrons were heavy, and averaged one crew per day. At this rate, a squadron would have lasted a fortnight without reinforcement, and it became common practise for the RAF commander on the island (AVM Sir Hugh Pughe Lloyd) to 'hijack' aircraft and their crews transiting Malta for the Far East. When the young Sgt Ivor Broom arrived at Malta, en route for Egypt and Malaya, he found himself immediately attached to No 105 Sqn on the island, and then to No 107 when the former unit was finally withdrawn to the UK long after its nominal five-week stint had been completed.

The unit was replaced by a large part of No 18 Sqn, which was on Malta from October until January 1942. Meanwhile, Sgt Broom flew missions from Malta for four months before being sent home with the two surviving original pilots from No 107. During that time No 107 Sqn sank 24 enemy ships at the cost of 24 crews. In the meantime, the loss of all the squadron's officer pilots had led to a field commission for Broom, who returned home a pilot officer. This was the first step on the ladder which eventually took Broom to the rank of air marshal. During his time on Malta, Malaya fell, and many of the Blenheim crews who did reach the Far East were lost, with a large of the survivors dying in appalling conditions in Japanese PoW camps.

The German air assault on Malta intensified in late 1941, and the island-based Blenheims began flying an increasing proportion of airfield attacks. On 4 January, for example, Nos 18 and 107 Sqns combined their ten serviceable aircraft and hit Castel Vetrano on the island of Sicily, destroying 30 enemy aircraft and scurrying away without loss. No 21 Sqn came back to Malta in January and stayed until late February, when the much depleted unit disbanded, reforming the same day back in the UK. No 107 Sqn was similarly reformed in England while No 18's survivors were absorbed into other MEAF Blenheim units. Other aircraft then took over the struggle from Malta, although the effect on Rommel had by then been largely achieved, it being reported that in late January, his Afrika Korps had been reduced to three days of supplies remaining!.

Operations from Malta did have their effect on Rommel's Afrika Korps, severely disrupting and dislocating his supply chain, and causing shortages which would eventually impede his fighting capability. It was against this background that Britain launched Operation *Crusader* in the Western Desert, with pre-emptive airstrikes beginning on 15 November 1941 – three days before the main assault. Blenheims, often heavily escorted, played their part, mounting close air support and airfield attack missions. Many No 113 Sqn aircraft were fitted with a long-barrelled 20 mm cannon firing forward through the nose glazing for ground strafing, and other units soon adopted similar modifications.

The Japanese attacks on Pearl Harbor and Malaya in December 1941 led to a rapid diversion of forces from North Africa to the Far East, and the Desert Air Force lost four of its nine Blenheim squadrons, with Nos

The bomb-carrying kangaroo on the nose of this Blenheim IV might suggest that the aircraft and aircrew pictured belonged to No 454 Sqn, an RAAF unit which briefly used Blenheims (principally Mk Vs) while working up, prior to beginning operations using Baltimores. Alternatively, another Aussie unit which used Blenheims was No 459 Sqn, although its Mk IVs lasted for just four months – the squadron's first operational type had been the Hudson (*via Bruce Robertson*)

Desert Air Force Blenheims were often extremely anonymous in respect to squadron markings, lacking codes and often individual letters too. This aircraft is known to have served with Nos 11 and 14 Sqns before being lost on 14 December 1941. Vokes filters on the bottoms of the engine cowlings prevented dust and sand from being sucked into the carburettors (*via Phil Jarrett*)

45, 84, 113 and 211 departing for Sumatra and Burma in January and February 1942, leaving behind Nos 8, 11, 14 and 55 Sqns, and the Free French *Lorraine* unit.

It was in the Middle East and North Africa that the Blenheim found itself being operated by Allied air force units under RAF command. Free French, Greek and South African units all operated Blenheims in a variety of roles.

Blenheims formed one flight of *Groupe Mixte de Combat* No 1 which formed at Odiham in August 1940 for service in Chad, this unit later merging with the Blenheim-equipped *Escadron Topic* at Maidugur to form *Groupe Réservé de*

Dated 29 April 1941, this photo appears to show an RAF Blenheim IV with two local levies, probably Transjordanian Bedouins. The Blenheim IV enjoyed some success in the Desert Air War, although it proved to be extremely vulnerable to enemy fighters (*via Aeroplane Monthly*)

Bombardement at Fort Lamy, Chad, in December 1940. This unit disbanded in March 1941, but formed the basis of Free French Flights Khartoum (at Gordon's Tree) and Sudan (at El Fasher) *Groupe Lorraine*, which operated alongside the RAF's No 1430 Flt, with support from No 47 Sqn. These units were absorbed when *Groupe Lorraine* formed in December 1941, and whose flights took the identities 'Metz' and 'Nancy', before becoming formally constituted as *Escadrille Metz* and *Escadrille Nancy* in November 1941.

After flying operations with Blenheim IVs, and latterly with Blenheim Vs, the unit embarked for the UK in October 1942, becoming No 342 Sqn, equipped with Handley Page Halifaxes within RAF Bomber Command. A second Blenheim-equipped Free French unit was *Groupe Bretagne*, although less is known of this unit's history and movements. Blenheims also served with the Free French Flying School at Bangui, and with *Groupe de Chasse* No 1 'Alsace'. The Blenheim served on with Free French forces, although not in the frontline, and there is photographic

This pair of Blenheim IVs were used by Free French forces during July 1941. The rearmost aircraft is a gun-pack-equipped Mk IVF, something of a rarity among the Free French Blenheims (*via Aeroplane*)

evidence of French Blenheim Vs in use as late as 1944 or 1945.

A number of nations bought Blenheims from Bristol before the war, including Finland, Romania, Turkey and Yugoslavia. Greece examined the aircraft, but did not order it before war broke out, when exports virtually halted. However, when Greece did order 12 Blenheim IVs after the conflict had started, these were delivered, and equipped No 32 Sqn. Subsequently, in March 1941 six ex-RAF Blenheim Is were passed to the Royal Hellenic Air Force as attrition replacements. Unfortunately all surviving Greek Blenheims were captured during the German invasion and occupation.

A Desert Air Force Blenheim IV flies over the wreckage of a Ju 52. Blenheims did gain a handful of air-to-air victories, and the lumbering Ju 52 represented a sitting duck – even for a Blenheim

Some Greek air force personnel escaped to Egypt, however, and fought on under RAF auspices, including the bulk of No 13 Sqn, with five Avro Ansons. The squadron formed part of the RAF's No 201 Group and was soon re-equipped with Blenheim IVs, and later with Blenheim Vs, which were used primarily in the maritime patrol role. From 1943 Blenheims gave way to Martin Baltimores.

Although the South African Air Force (SAAF) evaluated a Blenheim I before the war, it did not place an order for the type. This sole Blenheim I was delivered in December 1938 and was extensively evaluated by No 31 Sqn in the coastal patrol role, and was used in the search for the *Graf Spee*. The aircraft later shadowed two Italian ships as they raced for neutral Mozambique when Italy declared war, before attacking one of the vessels (the *Timavo*) with guns and bombs and forcing it to run aground. Unfortunately, the ship was refloated that night, and made its escape. Despite this active service, the type was judged to be unsuitable for SAAF

A Free French Blenheim IV taxies past a neat row of bombs, presumably waiting to be loaded aboard another of the squadron's aircraft. The cross of Lorraine replaced the fuselage roundel on this aircraft, and was also carried (alongside *Armée de l'Air* roundels) above and below the wings. Rudder stripes are also visible. (*via Peter H T Green*)

The *Armée de l'Air* enthusiastically embraced the Desert Air Force's loose approach to uniformity of clothing, and have even taken on charge the regulation squadron dogs. Most of the aircrew wear a mix of French and British issue clothing (*via Aeroplane*)

A No 14 Sqn Blenheim IV strafing enemy motor transport near Solumn on the north African coast on 17 June 1941. Many Desert Air Force Blenheim squadrons fitted 20 mm cannon for ground strafing. Without such a weapon, the Blenheim had to approach its targets far too closely, bringing it within easy reach of enemy small arms fire (*via Andrew Thomas*)

purposes, and the battle-weary Blenheim I was duly returned to the UK!

The SAAF's next connection with the Blenheim began in February 1942 when No 15 Sqn traded its Marylands for Blenheim IVFs whilst serving as part of the RAF's No 201 Group. These aircraft were equipped with underfuselage gun packs, and some also had a fixed forward firing 20 mm cannon shooting through the nose glazing. A detachment was made to Kufra oasis in April 1942 to prevent its use by the Germans, but the three aircraft got lost in the desert resulting in the deaths of 14 of the 15 men aboard through dehydration. These aircraft were abandoned at the crash site and were rediscovered in 1959, intact but for the fabric covering of their control surfaces. Armament was then removed and the aircraft were again abandoned – they were still present during the 1960s.

No 15 Sqn re-equipped with Blenheim Vs (inevitably referred to as Bisleys by the SAAF) in July 1942, moving base to Mariut, from where it began anti-ship and maritime patrol operations, although the unit continued to conduct some close support sorties. Highlights of the squadron's operations included participation in a massive convoy attack on 26 October 1942, during which three of the unit's aircraft were shot down, and three pilots received DFCs. On 17 February 1943 A No 15 Sqn aircraft bombed the U-205 after it had been depth-charged by HMS *Paladin*, prompting the crew's surrender. The unit finally traded its Bisleys for Bostons in July 1943.

Two further SAAF squadrons flew the Blenheim V in the anti-submarine role in the Middle East – No 16 Sqn received aircraft in November 1942, and was operational over the Indian Ocean from May until June 1943, when it converted to the Beaufort. No 17 Sqn at Little Aden flew the type between January and May 1943, when it received Hudsons. The Blenheim was also briefly used by No 60 Sqn, SAAF, in the reconnaissance role, and perhaps for an even shorter time by No 35 Sqn in the maritime patrol tasking.

The Blenheim IV began to disappear from frontline service during the summer of 1942, with some units converting to the Baltimore and Marauder (Nos 55 and 14, respectively) and others (No 15 SAAF and the Free French) to the Blenheim V. An exception to this process was No 11 Squadron, which continued with its ageing Blenheim IVs into August 1943.

The Blenheim had all but disappeared from the Desert Air Force's frontline by the time the Battle of El Alamein took place in October 1942, serving only with No 15 (SAAF) and No 11 Sqns, and with a number of coastal patrol and maritime reconnaissance units further removed from the frontline.

But as Blenheim operations 'wound down' at one end of North Africa, the type commenced combat at the other. The hopelessly outclassed Blenheim V, unwanted by Bomber Command, formed the backbone of the army support force committed to the Anglo-American invasion of North West Africa, codenamed Operation *Torch*. No 326 Wing, with Nos 13, 18, 114 and 614 Sqns re-equipped in England but having seen no action with their new mounts, moved to Algeria to support the Anglo-French invasion of what was primarily Vichy-controlled North Africa. But by 1943 the Blenheim (and perhaps especially the under-powered and overweight Mk V) was verging on obsolescence, and the scene was set for more hopeless heroics.

The aftermath of *Torch* was to provide the Blenheim force with its third and final VC, which was awarded to Acting Wg Cdr Hugh Gordon Malcolm, OC No 18 Sqn. The *London Gazette* printed the following VC citation;

'This Officer commanded a squadron of light bombers in North Africa. Throughout his service in that theatre his leadership, skill and daring were of the highest order. On 17th November, 1942, he was detailed to carry out a low-level formation attack on Bizerta airfield, taking advantage of cloud cover. Twenty miles from the target the sky became clear, but Wg Cdr Malcolm carried on, knowing well the danger of proceeding without a fighter escort. Despite fierce opposition, all bombs were dropped within the airfield perimeter. A Junkers 52 and a Messerschmitt 109 were shot down; many dispersed enemy aircraft were raked by machine-gun fire. Weather conditions became extremely unfavourable and as a result, two of his aircraft were lost by collision; another was forced down by enemy fighters.

'It was due to this officer's skilful and resolute leadership that the remaining aircraft returned safely to base. On 28 November 1942, he again led his squadron against Bizerta airfield which was bombed from a low altitude. The airfield on this occasion was heavily defended and intense and accurate anti-aircraft fire was met. Nevertheless,

An Army officer is helped on with his parachute harness, prior to a flight in a Blenheim of the Western Desert Communications Flight. The Blenheim was a useful light communications aircraft, with its relatively high speed and defensive armament. This aircraft was eventually abandoned at Landing Ground 04 on 28 June 1942 (*via Bruce Robertson*)

The Blenheim V offered no real improvement over the Mk IV, despite its uprated engines, more streamlined nose and fully retractable main gear. Even in North Africa, the aircraft was soon relegated to night operations, and to lower-risk activities like coastal patrols (*via Aeroplane*)

Blenheim V of No 114 Sqn is bombed up at Canrobert airfield, in Algeria, in December 1942 (*via Phil Jarrett*)

after his squadron had released their bombs, Wg Cdr Malcolm led them back again and again to attack the airfield with machine-gun fire. These were typical of every sortie undertaken by this gallant officer; each attack was pressed to an effective conclusion however difficult the task and however formidable the opposition.

'Finally, on 4th December 1942, Wg Cdr Malcolm, having been detailed to give close support to the First Army, received an urgent request to attack an enemy fighter airfield near Chouigui.'

As he walked out to the aircraft, Malcolm saw the OC of No 322 Wing responsible for fighter operations, Grp Capt Petrus 'Dutch' Hugo, and asked the South African (a 22-kill ace – see *Osprey Aircraft of the Aces 16 - Spitfire Mk V Aces* for more details)) for an escort. However, upon learning that no fighters could be spared, but knowing full well that a 'sweep' was in progress, Malcolm's reply is recorded as having been the single word 'Bullshit!' The citation continued;

'Wg Cdr Malcolm knew that to attack such an objective without a fighter escort – which could not be arranged in the time available – would be to court certain disaster; but believing the attack to be necessary for the success of the Army's operations, his duty was clear. He decided to attack.'

Of 11 available No 326 Wing Blenheim Vs forward-deployed to Souk el Arba, one aircraft burst the tyre on its tailwheel and slewed off the runway, while one more (from No 614 Sqn) developed engine trouble and crash-landed 15 miles from the landing ground. This left Malcolm's own aircraft, and four each from Nos 13 and 614 Sqns, although all but two crews were from No 18 Sqn. Although the target (an airstrip near Chouigui) had been attacked that morning, Malcolm was, according to eyewitness accounts, unable to see it as he circled the map reference he had been given (a hill). The aircraft then came under fire from German

flak, and from what were initially estimated as 26 Bf 109s. The citation again;

'He took off with his squadron and reached the target unmolested, but when he had successfully attacked it, his squadron was intercepted by an overwhelming force of enemy fighters .'

More Messerschmitts fell upon the Blenheims as they jettisoned their bombs (contrary to the citation's statement that the target was 'successfully attacked', and the formation began the long fight home.

'Wg Cdr Malcolm fought back, controlling his hard-pressed squadron and attempting to maintain formation. One by one his aircraft were shot down until only his own aircraft remained.'

Three of the Blenheim Vs actually made it back to Allied territory, only to be shot down. Wg Cdr Malcolm's was indeed the last aircraft to fall, some 30 miles west of the target, and unfortunately he and his crew were killed.

'In the end he, too, was shot down in flames. Wg Cdr Malcolm's last exploit was the finest example of the valour and unswerving devotion to duty which he constantly displayed.'

On the other two aircraft which made it back to Allied lines, everyone survived, although four of the surviving aircrew were injured. The last No 614 Sqn aircraft had already been damaged by enemy fighters, and it was written off in the subsequent forced landing at Souk al Khamis. The award of a VC to Malcolm recognised his personal bravery, and the spirit of the remaining crews, who had pressed on in the face of daunting odds. But while it turned the mission into a Pyrrhic

No 244 Sqn was the last frontline operator of the Blenheim V in the Mediterranean and Middle East theatre, flying anti-submarine patrols from Sharjah and Masirah until April 1944, when the type finally gave way to the Wellington. This aircraft was photographed, with members of the squadron in late 1942 (*via Andrew Thomas*)

Stripped of its dorsal turret and undernose guns, this Blenheim V (BA297) was used in the communications role until being struck off charge in April 1944. Photographed at Fayed, Egypt, on 3 April 1943, the aircraft has a non-standard D/F loop antenna above the forward fuselage (*via Bruce Robertson*)

This No 244 Sqn Blenheim V was written off after crashing on take-off at Salalah (Ras al Hadd), in Oman, on 26 November 1942. After being stripped for spares, the aircraft was dumped, leaving these two Wellington aircrew to find it two years later

victory, it could not camouflage the fact that the raid had failed utterly, with the Blenheims unable to find the target, and then being hacked from the sky with horrifying ease. The lack of fighter cover was of course a factor in the disaster, but at the end of the day, the failure simply underlined that the Blenheim V was too slow, too unwieldy, too poorly protected and too poorly armed to survive over a modern battlefield. It was almost as though nothing at all had been learned during the Battle of France some two-and-a-half years before.

No 18 Sqn was out of action until 18 December, with all aircrew sent to Sétif for recuperation. When they returned, the Blenheim V was relegated to night missions only. In January 1943 No 18 Sqn handed its Blenheim Vs to No 614 Sqn to await re-equipment, but subsequently took them back to fly more operations from 20 January! By the spring of 1943 the Blenheim was completely obsolete, and it was widely recognised that the aircraft was hopelessly vulnerable to enemy air opposition. Accordingly, the last close-support squadrons transitioned to the Boston, or adopted the less arduous coastal patrol role, while anti-shipping duties were passed on to the Beaufort.

This left four coastal reconnaissance and ASW units operating the Blenheim in the Middle East, with another squadron (No 162) including some Blenheims among its varied establishment for its calibration and Elint role. The last unit to cease bomber operations with the Blenheim V was No 614 Sqn, which did so in December 1943. No 162 discarded its last Blenheims in January 1944, as did No 8, while No 244 soldiered on at Masirah until April 1944. This marked the end for the Blenheim in frontline squadron service.

Even after this, the Blenheim hung on in the Middle East, flying in an assortment of second-line roles until war's end, when the last examples were finally scrapped. The final units equipped with Blenheims were Nos 75 and 79 OTUs, which disbanded in June and July 1945 respectively. The various Lend-Lease types (including the Bostons, Marauders and Mitchells, and the Hudsons and Venturas that had replaced Blenheims) were returned to the USA when the war ended, but there was no further need for the Blenheim, since Mosquitos, Beaufighters and Wellingtons were available in huge numbers.

BLENHEIMS EAST OF SUEZ

R AF squadrons in West Asia and the Far East had traditionally been engaged primarily in Colonial Policing operations, and as such were often the last units to re-equip with modern aircraft types. Bombing recalcitrant tribesmen on the North West Frontier did not require the latest aircraft types, nor did flag-waving in Singapore.

As it became increasingly clear that there was a real danger of war with Japan in the Far East, the RAF's squadrons in the area underwent a rapid and far-reaching modernisation programme. By late 1939, three Indian-based bomber units (Nos 11, 39 and 60) had re-equipped with Blenheim Is, and two more Blenheim squadrons (Nos 34 and 62) moved out to Singapore from the UK. When the war with Japan began, there were four Blenheim units in Malaya and Singapore – No 27 Sqn, with gunpack-equipped Mk IFs, and Nos 34, 62 and a detachment from No 60 with bomber-roled Mk Is. The latter detachment was only in Malaya for practise bombing, but was destined to be thrown into the battle which ensued alongside the other units. No 39 Sqn and the rest of No 60 Sqn were based in India, while No 11 Sqn had been transferred to the Middle East.

When Japan invaded Malaya on 8 December 1941, three Blenheim squadrons were in the very frontline, at Sungei Patani, Tengah and Alor Star, with the fourth (No 60) halfway down the Malay coast at Kuantan. The Japanese attack was three-pronged, with simultaneous landings at Singora (Siam) and at Kota Bahru, and strikes on Singapore city. No 27 Sqn took off at first light to attack the landing forces at Kota Bahru, but could not find their target in the very bad weather. Eight aircraft were lost while refuelling when Japanese bombers struck Sungei Patani. The remainder of the squadron withdrew to Butterworth the following day.

The last Blenheim VC to be promulgated (because records were lost until the war against Japan finished) was won by a pilot who participated

A Blenheim I of No 34 Sqn seen at Watton before departing for the Far East in August 1939. This aircraft was lost (still with No 34 Sqn) in January 1942 (via Bruce Robertson)

in the defence of Malaya. He was
Arthur Stewart King Scarf, less formally known as John, or within the
RAF as 'Pongo'. Scarf was a member
of No 62 Sqn, which moved to the
Far East from Cranfield in August
1939. In February 1941 the
squadron transferred to the
Malayan airfield of Alor Star, near
the border with neutral Siam, in
anticipation of Japanese expansion
through Siam to Malaya. This mate-
rialised on 8 December when Siam
surrendered, and Japanese aircraft
operating from Thai bases attacked
RAF airfields in Malaya. On that morning No 27 Sqn lost eight of its
twelve IFs as they refuelled at Sungei Patani following an abortive sortie,
withdrawing the survivors to Butterworth, where they were attacked
again, along with the Blenheims of the resident No 34 Sqn. Three of the
latter unit's aircraft had already been lost during a raid on Singora. No 60
Sqn's eight aircraft detachment at Kuantan flew a mission against Japanese invasion ships, losing two aircraft – the unit's four surviving
Blenheims pulled out to Tengah, Singapore, on 9 December.

Within two days, all the RAF's Blenheims had been driven from their
original bases, and the remnants of three squadrons were all concentrated
at Butterworth. Before this withdrawal, on 8 December No 62 Sqn's
Blenheims were caught on the ground at Alor Star, and many were
destroyed. The survivors moved south to Butterworth, joining the remnants of Nos 27, 34 and 60 Sqns. With the Japanese occupying Singora
and Sungei Patani airfields, the Blenheims were tasked with attacking the
airfields. Six Blenheims from No 34 Sqn (three of them flown by No 60
Sqn crews) duly attacked in the morning, losing three aircraft in the
process. The remaining crews were ordered to fly a second raid in the
afternoon, with all of No 62 Sqn's flyable Blenheims and pilots, including 'Pongo' Scarf, taking part. The *London Gazette* printed his VC citation (run here with additional information in italics);

'On 9th December 1941, all
available aircraft from the Royal Air
Force Station Butterworth, Malaya,
were ordered to make a daylight
attack on the advanced operational
base of the Japanese Air Force at Singora, Thailand. From this base, the
enemy fighter squadrons were supporting the landing operations

'The aircraft detailed for the sortie
were on the point of taking off when
the enemy made a combined dive-
bombing and low-level machine-
gun attack on the airfield. All our
aircraft were destroyed or damaged

Aircrew clamber aboard a Blenheim
I 'somewhere in Bengal'. They are
probably Indian aircrew, and the air-
craft was most likely one of the ex-
RAF Blenheim Is issued to No 3
Coast Defence Flight, which moved
from Calcutta to Bassein, in Burma.
The unit re-equipped with
Blenheims in mid-December 1941
(*via Bruce Robertson*)

Although of poor quality, this rare
shot of a No 39 Sqn Blenheim I
reveals the aircraft marked with the
winged bomb insignia of the unit on
the central white band of its fin
flash. The aircraft is seen in its camouflaged revetment in Aden during
operations against the Italians in
East Africa. No 39 Sqn had traditionally been based in India, although its
Blenheims were diverted to Aden
while en route from Singapore to
the Middle East
(*via Andrew Thomas*)

with the exception of the Blenheim (*L1134*) piloted by Sqn Ldr Scarf – *Scarf's crew were Flt Sgt Freddie Calder (observer) and Flt Sgt Cyril Rich (wireless operator/air gunner)*. This aircraft had become airborne a few seconds before the attack started.

'Sqn Ldr Scarf circled the airfield and witnessed the disaster. It would have been reasonable had he abandoned the projected operation which was intended to be a formation sortie. He decided, however, to press on to Singora in his single aircraft. Although he knew this individual action could not inflict much material damage on the enemy he, nevertheless, appreciated the moral effect which it would have on the remainder of the unit, who were helplessly watching their aircraft burning on the ground.'

En route to the target, Flt Sgt Rich demonstrated the Blenheim's ability to defend itself, beating off attacks by several Japanese fighters which made the mistake of attacking singly, in turn. Over the target, things got hotter, with further fighters starting to make simultaneous attacks.

'Sqn Ldr Scarf completed his attack successfully.' *Rich strafed rows of parked aircraft as Scarf dropped his bombs.* 'The opposition over the target was severe and included attacks by a considerable number of enemy fighters. In the course of these encounters, Sqn Ldr Scarf was mortally wounded.' *In fact, Scarf was seriously wounded, with a shattered left arm and bullets in his back, but the wounds may not necessarily have been fatal. When he later reached hospital, Doctors were not confident that they could save Scarf's arm, but were optimistic that his life would be saved.*

'The enemy continued to engage him in a running fight, which lasted until he regained the Malayan border. Sqn Ldr Scarf fought a brilliant evasive action in a valiant attempt to return to his base. Although he displayed the utmost gallantry and determination, he was, owing to his wounds, unable to accomplish this.' *Once the Japanese fighters had broken off their attacks, Rich came forward from his turret and supported Scarf in his seat, while Calder helped Scarf maintain his grip on the control column. It soon became clear that Scarf would pass out before they could reach Butter-worth, and the airmen decided to land at Alor Star instead.* 'He made a successful forced-landing at Alor Star without causing any injury to his crew.' *They landed wheels up, sliding through rice paddies and coming to a halt only about 100 yards from the hospital. Rich and Calder*

No 39 Sqn's Blenheim Is lined up at Kallang, Singapore, (with 'YH-N' of No 11 Sqn) during early 1940, before the unit left the Far East
(*via Peter H T Green*)

Blenheim Is of No 27 Sqn at Risalpur on 13 February 1941. That day the squadron moved to Kallang, Singapore, where is subsequently suffered serious losses in heavy fighting the following December
(*via Geoff Thomas*)

lifted Scarf from the cockpit and laid him on the port wing. All three airmen then lit up cigarettes, oblivious of the fuel splashing from the aircraft's ruptured tanks! 'He (Scarf) was received into hospital as soon as possible but died shortly after admission. Sqn Ldr Scarf displayed supreme heroism in the face of tremendous odds and his splendid example of self-sacrifice will long be remembered.'

In a terrible coincidence, Scarf's new wife Elizabeth was a nurse at the Alor Star hospital, and Scarf laughed and joked with her from his stretcher. She gave two pints of her blood during a never-completed transfusion, but Scarf died from secondary shock.

Scarf's sacrifice marked the end for the Blenheim in the battle for Malaysia, with all remaining aircraft and crews (about eight Mk Is and three Mk IVs) concentrating at Kallang, Singapore. On 9 December three of the six newly-arrived No 27 and No 60 Sqn crews were ordered to make a one-way mission against Singora, along with three crews from No 34 Sqn, where they attacked Japanese shipping. Briefed to 'land where they could', and with insufficient fuel to return to Singapore, five of the aircraft crashed in the jungle, one aircraft returning via Butterworth. On 23 December, No 60 Sqn handed over its remaining aircraft and some personnel to Nos 27, 34 and 62 Sqns. The rest of No 60 Sqn sailed for Rangoon. All Blenheim squadron records were lost in the confusion, and it was not until June 1946 that Scarf's VC was announced.

In Sumatra, the ex Malayan-based Blenheims joined No 84 Sqn, which had flown in from Egypt on 23 January. The allied forces in Sumatra came under constant attack, and seven Blenheims were lost on 7 February during an air raid. With the invasion of Singapore on 8 February, the RAF's surviving aircrew withdrew to Sumatra with their few surviving aircraft to continue the fight. The three Blenheim units could only muster about a dozen aircraft between them. Three days later, Sumatra came under threat, and the force from Singapore withdrew again to Java.

One of Sumatra's two airfields was taken by Japanese paratroops on 14 February, and the remaining six Blenheims withdrew to Kalidjati, on Java, on 18 February. From here they mounted attacks on shipping and the approaching invasion force, even despatching a submarine on 23 February. On 1 March, with Kalidjati about to fall, No 84 Sqn destroyed its last Blenheims and its men attempted to escape. Some succeeded by commandeering a lifeboat and making an epic 47-day journey to Australia.

But while the RAF's Blenheims were unable to prevent the loss of Malaya, Singapore and the East Indies, they did play a part in preventing the fall of India, and in the long campaign which eventually recaptured Burma. Seven frontline Blenheim bomber squadrons participated in the campaign, successively using the Blenheim I, the Blenheim IV and even the Mk V, or Bisley. As No 84 Sqn rushed from Heliopolis to Palembang, No 113 Sqn was sent from the Middle East to Mingaladon-Rangoon on 7 January, where it absorbed the rump of No 60 Sqn,

Blenheims of No 62 Sqn enjoying the calm before the storm, flying off the Malay coast shortly before the Japanese invasion. The code 'FX' was used only very briefly (*via Andrew Thomas*)

the rest of the unit having been expended in the Malayan campaign. While based at Mingaladon, the squadron raided Bangkok, dropping 11,000 lbs of bombs on Japanese forces there. In the next weeks, the squadron accounted for about 60 enemy aircraft in bombing and strafing attacks on Japanese-held airfields. But there was no stopping the Japanese juggernaut, and the campaign soon became a fighting retreat. At the end of February No 113 then moved back to Magwe with No 45 Sqn (another former MEAF unit), withdrawing as the Japanese continued advancing.

Attacks on Magwe during late March reduced Blenheim numbers, but the RAF did hit back, most notably on 21 March when a force of ten Hurricanes and nine Blenheims devastated their former base at Mingaladon. With the situation stabilised, the surviving Blenheims were withdrawn to Akyab, and then back into India, where the dispersed remnants of the squadrons were reconstituted and reformed.

The Japanese were not supermen, nor were they invincible. In fact, the Japanese Army's refusal to issue mosquito nets to its aircrew meant that many were debilitated by Malaria! Moreover, even the mighty Mitsubishi A6M Zero was only very lightly armed, lacked any armour, and was thus vulnerable to even light calibre return fire. Although more often than not Blenheims were hacked from the skies by Japanese fighters, there were occasions when Blenheim crews exacted their revenge. One such episode occurred on 22 May 1942. Wt Off Huggard of No 60 Sqn found that his Blenheim IV was the only one of three to reach its target (the newly-captured now-Japanese airfield at Akyab), which he attacked at low level, and then fled, at wave top height, over the Bay of Bengal.

The Japanese Army Air Force's elite 64th Sentai scrambled five Ki-43s to pursue the impudent raider. The Blenheim's cool-headed gunner, one Sgt Jock McLuckie, damaged the first aircraft to attack him on its first pass, forcing the 10-kill ace piloting it to return to base. Sgt Maj Yoshito Yasuda was soon joined by his wingman, Capt Masuzo Otani, who was also hit by a well-aimed burst from McLuckie, and who was similarly forced to return to base. The next formation

Barely discernable 'AD' codes identify these Blenheim IVs as belonging to No 113 Sqn. This rare post-invasion photograph was probably taken just before the final fall of Magwe, after which the squadron was dispersed (*via Geoff Thomas*)

A Blenheim IV of No 60 Sqn waits to be bombed-up at Asansol during 1942. From here, No 60 Sqn mounted a series of attacks on Japanese bases and forces in Burma, with some success (*via Peter H T Green*)

Even after it had been withdrawn from frontline service, the Blenheim had a role to play. This Mk V was used by an Airborne Surgical Unit, carrying personnel and equipment, including the mobile X-ray gear seen in this carefully posed propaganda photograph. The Bristol BX turret is shown to advantage (*via Bruce Robertson*)

The last Blenheim IVs lingered on until July 1943 with Nos 11 and 60 Sqns at Feni, the remaining units having converted to Mk Vs, which remained in service for a few weeks longer. This aircraft (Z7648) was serving with No 11 Sqn when it was photographed in April 1943 (*via Geoff Thomas*)

of Ki-43s to catch the Blenheim caught up with the aircraft half an hour after the attack, and was led by the famous 18-kill ace, Lt Col Tateo Kato, a veteran of the Sino-Japanese war and commander of the 64th Sentai (see *Osprey Aircraft of the Aces 13 - Japanese Army Aces 1937-45* for more details). McLuckie hit the colonel's Oscar as it pulled up from its first firing pass, setting the flimsy fighter alight. Japanese legend has it that Kato deliberately half rolled and dove into the sea, realising that he could not make it back to base. The truth is likely to have been more prosaic – that fire burned through the fighter's control wires, or that Kato himself had been badly hit and lost consciousness. Whatever the cause, Kato perished as his fighter hit the sea, and his wingmen broke off their attack and returned to base.

The campaign in Burma was a fighting retreat, and ended with the Japanese further forward than they had been when they started out. The next action in which the Blenheim was involved saw the British actually denying the Japanese forces territory. This was the defence of Ceylon, threatened with the possibility of a Pearl Harbor-type strike. Vice Admiral Nagumo, who had led the Pearl Harbour raid, actually led a carrier force into the Indian Ocean to attack Ceylon, but allied code-breakers gained early intelligence and the island's garrison was reinforced.

By the end of March 1942, the island was defended by 50 Hurricane fighters, a handful of FAA Fulmars, six Catalina flying boats and 14 Blenheim IVs of No 11 Squadron, together with a handful of FAA Albacores. This rag-tag force was able to beat off the Japanese air attacks, and in doing so, the Blenheims mounted a number of recce missions and one costly, but ultimately successful, air strike. On 9 April 11 Blenheims attacked Japanese ships without fighter escort. Two aircraft aborted, and five more were lost to Zeroes and AAA, but they hit many of their targets.

The end of 1942 saw the British and Indian armies making intensive preparations for a campaign which would hit back at the Japanese. Pre-

liminary probing operations began in December 1942, in preparation for an all-out offensive.

Blenheim IVs and Vs supported the Arakan campaign, spearheading the anti-shipping strikes off the island of Akyab, and bombing enemy airfields (most of which had once housed RAF Blenheims, including Magwe), troops, river barges and transport routes. Air opposition was less of a problem

than it had been in the earliest days of the war in the Far East, not least because Hurricanes usually served as escorts, rather than inadequate Buffaloes. On 9 September 1942 nine Blenheims from Nos 60 and 113 Sqns lost two aircraft to Zeroes during an anti-shipping strike, with three more having to force-land when they returned to base with major damage.

The Blenheims remained on the offensive as the 'Forgotten' 14th Army pushed eastward, operating from forward airfields. They then covered the Army as it withdrew back into India in May 1943 before the summer monsoon. Although little ground was gained, the offensive proved successful, demonstrating Japan's weaknesses, and showing that the Allies could counter an enemy which had once been regarded as invincible.

The five remaining Blenheim V squadrons all re-equipped with Hurricanes. Nos 11, 34 and 60 at Feni, Madras and Yellahanka all discarded Blenheims by the end of July, while No 113 at Feni flew its last sorties with the Bristol fighter-bomber on 15 August and No 42 lingered on at Kumbhirgram until October 1943. The Blenheims and their crews had fought long and hard as the Japanese advanced, helping halt the enemy before it could seriously threaten British India. The aircraft had then been used to carry the offensive back to the Japanese, and had supported the first faltering steps in the offensive which would eventually sweep the Imperial Japanese forces back to their island homeland, and final defeat.

Although the history books record that Mk Vs ceased frontline operations in India in October 1943, the type actually lingered on much longer. A handful of Blenheims were used in the calibration, meteorological and survey roles until mid-1944, and at least one flew a handful of offensive operations. According to RAF lore, the CO of Beaufighter-equipped No 176 Sqn found abandoned Mk V BA191 at Kangla and got it airworthy for an air test on 8 May 1944, using another aircraft discovered at Imphal as a spares source. The following day he flew the aircraft

(with navigator P G Bowen and a 'geordie' LAC acting as gunner) on a mission to bomb a Japanese field HQ and a bridge. Another sortie (against a bridge and MT) was made on 11 May. The last Blenheims in the Far East were retired in March 1945, or perhaps even when No 1 Air Gunnery School (India) disbanded in July 1945.

No 113 Sqn at Feni flew the Blenheim until August 1943, when it re-equipped with Hurricanes. One of the squadron's anonymous-looking Blenheim Vs is seen here. No 42 Sqn at Kumbhirgram lingered on until October, however. This was the last frontline Blenheim bomber squadron in the Far East, and was outlasted by No 614 in North Africa, which flew Mk Vs until December, and by No 244, which retained Blenheims (albeit in the maritime role) until May 1944 (*via Geoff Thomas*)

This Blenheim IV, captured by the Japanese in 1942, found its way into the hands of the Indonesian nationalists, who fitted it with 950 hp Nakajima Sakae engines and prepared it for use in the reconnaissance role. The bomber crashed after making one flight from Masupati in 1946 (*via Bruce Robertson*)

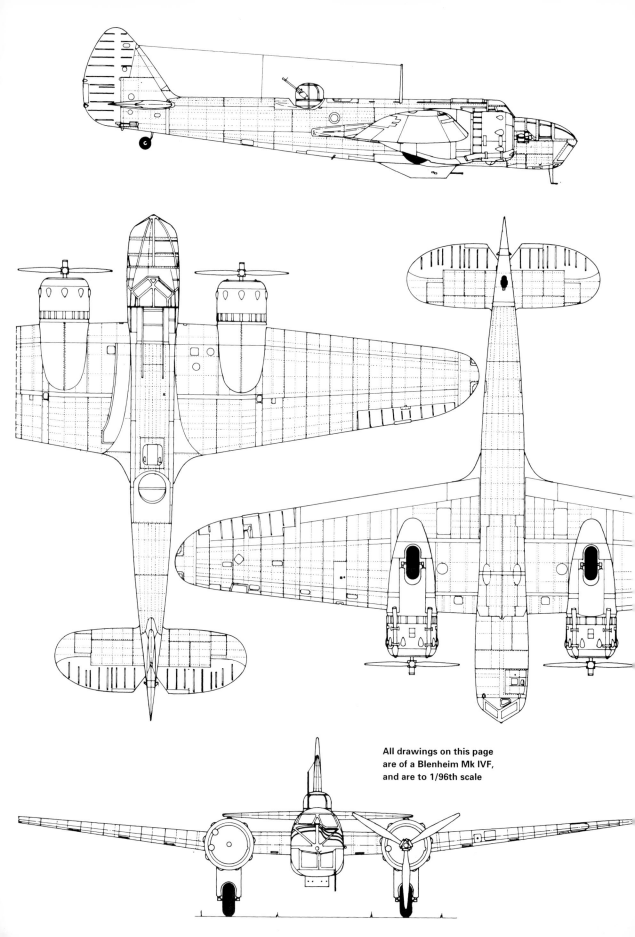

All drawings on this page
are of a Blenheim Mk IVF,
and are to 1/96th scale

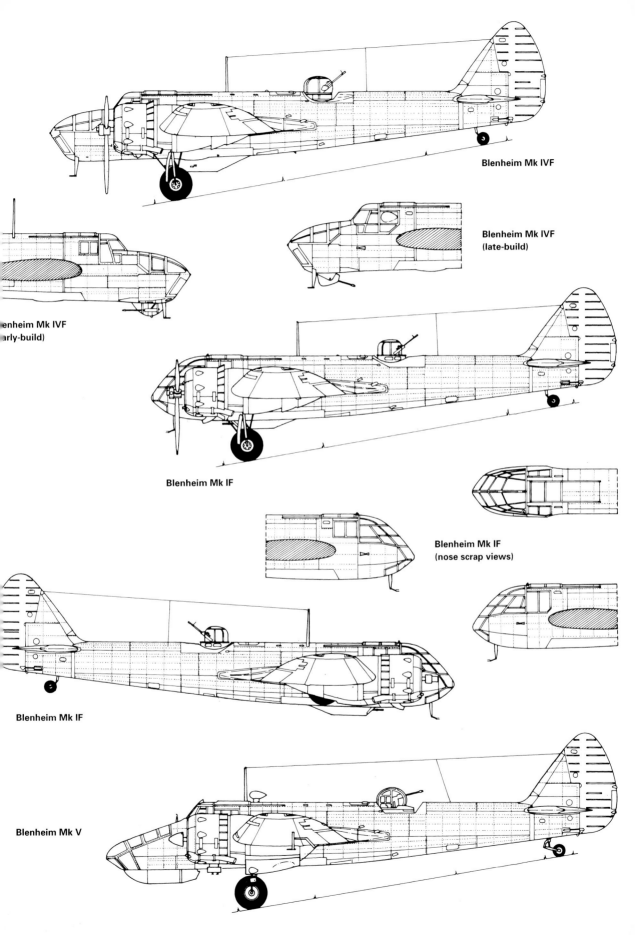

Blenheim Mk IVF

Blenheim Mk IVF
(late-build)

Blenheim Mk IVF
(early-build)

Blenheim Mk IF

Blenheim Mk IF
(nose scrap views)

Blenheim Mk IF

Blenheim Mk V

COLOUR PLATES

1

Blenheim V BA429/P of No 8 Sqn, Khormaksar, Aden, 1943

No 8 Sqn was one of the longest serving Blenheim operators, receiving Mk I bombers in April 1939 as replacements for its antique Vickers Vincents in the colonial policing role. The former aircraft were displaced from home-based Bomber Command units by the newer Mk IVs, but while they were cast-offs, they were most welcome in the RAF's Near-East Air Force (NEAF). The Blenheim Is served with No 8 until October 1941 (alongside Swordfishes and Marylands), although the Blenheim IV (first received in January 1941) soldiered on until August 1942, when it was replaced by the Mk V. This variant in turn served until January 1944, latterly alongside Hudsons and Wellingtons. Although initially assigned a bomber role, No 8 Sqn had switched to anti-submarine patrols and internal security duties after the Italian surrender in East Africa. These were the only roles carried out by the squadron's Blenheim Vs, which wore a variety of colour schemes. This aircraft (which served only briefly with No.8 Sqn, before flying with No 13 'Hellenic' and No 16 SAAF Sqns) wears typical early European Coastal Command colours, with Sky undersides and Extra Dark Sea Grey and and Dark Slate Grey topsides. The aircraft boasts a C-Type fin flash and a C1 fuselage roundel. Other No 8 Sqn Mk Vs wore the grey and white scheme shown in profile 36.

2

Blenheim IV Z6089/F-OO of No 13 Sqn, Odiham, Dieppe Raid, 19 August 1942

By the time of the Dieppe raid, the Blenheim had virtually disappeared from No 2 Group, and the action marked the type's swansong in Europe, with the remaining units converting to the Blenheim V for service in North Africa. No.13 Sqn mounted two waves during the Dieppe raid, dropping 100-lb phosphorous bombs on the landward side of the coastal guns while the Blenheim IVs of No 614 Sqn streamed smoke. This aircraft was written off as a result of battle damage sustained in the raid. By the time of Dieppe, some Army Co-operation Command Blenheims had dark green and ocean grey upper surfaces, while others used medium sea grey instead of ocean grey. A number of aircraft also wore black undersurfaces. This machine has a more dated appearance, though, with its light undersides, and dark green and dark earth topsides. Nevertheless, it has the later twin-Browning turret installation, but unusually retains the early glazed undernose turret.

3

Blenheim V EH333/L of No 13 Sqn, Blida, Algeria, June 1943

In North Africa, the Blenheim Vs of No 326 Wing did not carry squadron code letters, and were frequently swapped between squadrons. This aircraft wears the standard 'North African' Blenheim colour scheme of azure undersides and earth/mid stone topsides. Vokes filters are fitted below the engine nacelles. Heavy losses soon led to the Blenheim Vs being restricted to night operations, but they never received black undersides. For the last few months of its existence as a Blenheim unit, No 13 Sqn operated primarily in the coastal patrol and convoy escort missions. This aircraft crashed at Blida on 19 July 1943.

4

Blenheim IV L9192/Q-WV of No 18 Sqn, Great Massingham, March 1941

This No 18 Sqn Blenheim is typical of those which participated in No 2 Group's near-suicidal campaign of anti-shipping 'Roadsteads' and 'Channel Stops', and in the heavily escorted and deliberately provocative 'Circuses'. Camouflage demarcation lines and the presentation of squadron codes were never really standardised, and this aircraft's two-letter 'WV' codes aft of roundel are particularly squat, while the demarcation between upper and lower surfaces is rather high. The aircraft combines the later twin-Browning midupper turret with the early glazed undernose turret, but unusually, is fitted with propellor spinners.

5

Blenheim V BA875/W of No 18 Sqn, Canrobert, Tunisia, December 1942.

The subject of Iain Wyllie's cover painting, this Blenheim V was the aircraft flown by Wg Cdr H G Malcolm on the 4 December daylight raid during which he won the Victoria Cross. No 18 Sqn formed part of No 326 Wing, along with the similarly equipped Nos 13, 114 and 614 Sqns. Aircraft wore single letter codes, with no squadron prefixes, this machine being painted in the standard Blenheim V colour scheme for North Africa. The red code 'W' was applied forward of the roundel, with nothing aft. The aircraft is fitted with both the standard Mk V two-gun mid-upper turret and undernose turret.

6

Blenheim IF K7090/ZK-V of No 25 Sqn, North Weald, February 1940

From September 1938 RAF fighters were painted with black and white undersides, with black below the port wing, and usually divided along the centreline or the port wing root. Ailerons were sometimes left in the original colour, and some aircraft even had a black aileron on the white wing! Some units moved the fuselage roundel aft on the port side to make room for the codes, others applied the individual letter forward on both sides, whilst still others painted the codes over the trailing edge fillet. This aircraft wears an A type fuselage roundel – the yellow ring was not added around it until May-June 1940. The individual aircraft letter 'V' was repeated in white on a painted panel below the windscreen. K7090 was one of No 144 Sqn's original bombers, serving as a fighter with No 248 Sqn before joining No 25 Sqn on 22 February 1940 – the aircraft may have been one of those used by the unit for radar trials. By that time the Blenheim was fitted with a four-gun ventral gun pack. The aircraft passed to No 55 OTU on 21 November and then to No 54 OTU after a month. It was lost in a fatal accident while orbiting a beacon just north of Church Fenton on 18 July 1941.

7
Blenheim IF L1327/I-RO of No 29 Sqn, Digby, June 1940

Looking very much like a standard Blenheim bomber rather than a nightfighter (apart from its ventral gun pack), this aircraft of No 29 Sqn wore Sky undersides, with green/dark earth topsides. Sky was devised by the company Titanine, and was a similar shade to the Duck Egg Green adopted by Blenheim bombers just before the war. It was originally known as Camotint, and gave an exceptionally smooth finish. Sky was officially adopted in June 1940. A full height fin flash occupied most of fin, and L1327 had an A1 type roundel on the fuselage. Unusually, the aircraft had its serial repeated in its original position on the rudder. Nightfighters only really started to look the part when they began to receive soot black (RDM2) overall from December 1940, and dark areas of their national insignia were often overpainted from mid-1941, while unit codes were repainted red in the autumn of 1941.

8
Blenheim IV R3612/V-BL of No 40 Sqn, Wyton, July 1940

By the end of the Battle of France, most No 2 Group Blenheims looked very much like this. Light (usually Duck Egg Green, but sometimes, from June 1940, sky) undersides would soon start to give way to black, however. RAF aircraft operating in, or over, France began to receive tricoloured fin flashes in March 1940 in an effort to bring them into line with *Armée de l'Air* machines, which had rudder stripes. The RAF was nervous about applying extra paint to control surfaces, so the tail stripes were applied to the fin, rather than the rudder. The new full height fin flashes varied in width, and in inclination, some being applied to be vertical with the aircraft sitting on the ground. Several squadrons followed the practise of outlining the individual aircraft code letter in white, and some painted the letter itself in the flight colour, or in red.

9
Blenheim V AZ942 (formerly code letter S) of No 42 Sqn, Kumbhirgram, India, late 1943

As the war in the Pacific progressed, it became increasingly clear that the RAF's national insignia could, under certain circumstances, show up as simple red discs – too similar by far to the Hinomaru worn by Imperial Japanese aircraft. Red was accordingly dropped from roundels and fin flashes during 1942, resulting in what were known as SEAC roundels and fin flashes in blue and white only. The Blenheims weathered extremely rapidly in the harsh monsoon climate of India, and most Blenheim Vs looked considerably less pristine than this machine. No 42 Sqn used the Mk V from February to October 1943, but this aircraft crashed near Jessore on 25 August 1943.

10
Blenheim I L8479/W-OB of No 45 Sqn, LG 08/Mersah Matruh, Egypt, September 1939

This Blenheim I is one of those repainted in desert camouflage shortly before an inspection of the squadron by King Farouk in September 1939. The aircraft retained the standard Night Black undersides of the period, with Dark Earth and tan topsides, the latter colour replacing the normal green. No fin flash was applied at this early date, and the aircraft had an A-type fuselage roundel. Squadron codes were applied in the uppersurface camouflage colours of tan and earth contrasting with the base colour, tan on earth and earth on tan. Some of the normally transparent nose panels were also over-painted. The aircraft was fitted with external bomb racks.

11
Blenheim IV (serial unknown)/K-OB of No 45 Sqn, Muqueibila, Palestine, July 1941

No 45 Sqn converted to the Blenheim IV in March 1941, before moving to the Far East in February 1942. This aircraft is typical of tropicalised Blenheim IVs, with Vokes filters over the carburettor inlets on the bottom of each engine cowl. The desert environment wreaked havoc on the bungee cords which opened and closed the bomb doors, and aircraft frequently flew with the latter removed, as shown here.

12
Blenheim IV L4841/N-TE of No 53 Sqn, Odiham, 1939

The first Blenheim IVs were delivered to this Army Co-operation Command squadron wearing factory applied night, dark earth and dark green camouflage, with very hard and 'neat' demarcation between the colours. The aircraft soon had their fuselage roundels converted to B-type, and 'TE' codes were applied aft of the roundel, with individual letters forward of the national marking. When identity codes were applied, the aircraft lost their squadron badges, which were applied, very small, in a 'grenade'-type frame on the tailfin. Serials were carried on the rear fuselage, and were repeated on the rudder. The basic mid-upper turret was fitted, with a single Vickers Gas Operated Type-K gun. There was no undernose turret fitted.

13
Blenheim IV Z9601/H of No 55 Sqn, Fuka, Western Desert (Egypt), early 1942

When No 55 Sqn resumed operations in North Africa in September 1941, it mainly flew anti-shipping sweeps, and as such experimented with colour schemes optimised for overwater operations. This Blenheim IV wore a decidedly unusual colour scheme, with Azure undersides and dark Mediterranean blue/Mediterranean blue topsides. The aircraft had a rectangular fin flash with a broad white central stripe and an A1 fuselage roundel. It was identified within the unit by a dirty white code (H) forward of the roundel, with nothing aft. The aircraft was fitted with the later two gun mid-upper turret and the solid two-gun undernose turret. The camouflage was weathered and faded, with the new topside colours applied crudely over the original earth/green camouflage. The aircraft's leading edges were also very worn.

14
Blenheim I L1145 of No 57 Sqn, Upper Heyford, 1938

The Blenheim IV was designed for Army Co-operation Command, and its squadrons took the first examples to leave the production line. However, No 57 Sqn moved to France in 1939 *before* it could convert to the new variant, and thus initially retained Blenheim Is. This aircraft wears the standard camouflage scheme applied at the factory, with Night undersides, dark earth and dark green topsides. The aircraft has no fin flash and wears an A1 fuselage roundel but no squadron codes. The serial was carried on the rear fuselage, and was repeated on the fin. Following the Munich Crisis, some squadrons had no two-letter codes to adopt in place of their numerical squadron identities, and thus flew around almost unmarked. The adoption of B-type roundels was far from universal, with several stations retaining the pre-Munich A1 roundel for many months. This aircraft has the standard early single K-gun type mid-upper turret, which is shown here in the retracted position. The unit badge is carried in a traditional 'grenade' on the lower fin.

15
Blenheim I L1134/F-PT of No 62 Sqn, Alor Star, Malaya, 9 December 1941

Having been hastily delivered to the Far East from UK stocks, FEAF's early Blenheim Is wore much the same colour scheme as the UK-based aircraft, with Night black undersides, dark earth/dark green topsides and the usual national insignia for the period, including the underwing roundels and equal-width, almost square, fin flash promulgated in late 1940. The aircraft was fitted with new (locally fitted) Vokes filters below the cowlings – no-one had pre-painted these to match the normal undersurface colour, so they remained grey. This was the aircraft in which Sqn Ldr Scarf won his VC.

16
Blenheim IVF Z5722/Z-WM of No 68 Sqn, High Ercall, 1941

This Blenheim IVF was the personal aircraft of Wg Cdr the Honourable Max Aitken, OC No 68 Sqn, in preference to one of the Beaufighters with which the squadron was then re-equipping. The aircraft survived as a squadron 'hack' after the Beaufighters were completely sorted, but was lost at Coltishall on 6 June 1942 after a pilot attempted to take off without closing the cooling gills, and hit a flagpole. Aitken, son of Lord Beaverbrook, had flown Blenheim Is with No 601 Sqn on the first day of the war, and after commanding No 68 Sqn and the Banff Strike Wing, managed to fly a Mosquito sortie on the last day of the war, too – he racked up 15 air-to-air victories during the conflict. Finished in Smooth Night overall, with night-bomber type national insignia and markings, the aircraft had the later rectangular fin flash with equal stripes, an A1 fuselage roundel, medium sea grey codes and a red serial. This Blenheim had the early single gun mid-

upper turret and no undernose turret, but was fitted with the forward-firing ventral gun pack with four 0.303-in Brownings. The aircraft is equiped with AI Mk III radar. The rather rough and draggy ultra-matt Special Night was replaced by smooth night on night-fighters and intruders in early 1942, the new paint having a more glossy finish. A very small number of Blenheim bombers eventually wore an overall black finish, these being aircraft used exclusively in the night intruder and night-bombing role at the very end of the Blenheim's career with No 2 Group – a handful of these aircraft may even have worn the red code letters associated with late-war night bombers. This colour scheme was chosen for the second Blenheim/ Bolingbroke restoration by Graham Warner, after the first (in No 105 Sqn colours) had crashed in 1987.

17
Blenheim IV P6915/A-UX of No 82 Sqn, Watton, early 1940

Struck off charge following damage incurred by intercepting Bf 109s during an armed recce on Abbeville raid on 7 June 1940, this No 82 Sqn Blenheim IV still boasted a glazed undernose gun turret this late into the conflict. The pale undersides were Duck Egg Green rather than the later Sky. Interestingly, the aircraft appeared to have a less distinct demarcation between its dark green and dark earth topside colours. This particular aeroplane had a full height fin flash, whose central broad white stripe occupied virtually the whole of the vertical fin. A famous photo of the aircraft shows other machines with a wide variety of fin flash styles, some with narrow white stripes, some broad, some full height, some short, and some with no fin flash at all. The aircraft wore an A1 roundel on the fuselage, whose yellow outer ring overlapped the squadron codes.

18
Blenheim IV T1828/V-UX of No 82 Sqn, Bodney, February 1941

By 1941 most Blenheim operations were conducted in daylight either as part of heavily escorted 'Circuses', intended to draw up enemy fighters, or as suicidal anti-ship attacks. The Blenheims were also used for night intruder sorties and to mount diversionary raids in support of Main Force operations. For the latter sorties, squadrons tried to use the small number of Blenheims with black undersides, painted with a washable matt black distemper – the number of aircraft with black undersides increased as night operations assumed greater importance. This aircraft was unusual in having a high demarcation line, and in having a black fin and rudder. The Night black undersides extended almost to the roof. Blenheims used principally at night tended to be fitted with 'stickleback'-type flame-damper exhausts (not worn by T1828).

19
Blenheim I L1381/G-VA of No 84 Sqn, Menidi/Tatoi, Greece, April 1941

Many reference books have repeated the erroneous belief that the codes 'VA' were allocated to the Blenheims of No 113 Sqn during the Greek campaign, when in fact these were the letters used by No 84 Sqn. Many photos of the latter unit's Blenheims have thus been mis-identified. Most of the Blenheims which saw service in Greece were Mk Is. At the end of August 1940 the squadron moved from Shaibah, in Iraq, to Heliopolis, in Egypt, with 'B' and 'C' Flights moving on to Fuka in the Western Desert. Five weeks later, on 2 October 1940, 'A' Flight flew to Greece (via Crete), arriving at Elevsis and then moving to Menidi (Tatoi). Here the remaining flights caught up, and the squadron settled in. No 84 began to replace its Mk Is with Mk IVs, and these proved better able to operate from Menidi's short, poplar-line, runway! This aircraft is unusual among No 84 Sqn's Blenheim Is used in Greece in that most retained Western Desert camouflage, with the normal green over-painted in tan.

20
Blenheim IV Z7427/K-RH of No 88 Sqn, Attlebridge, August 1941
Even as late as August 1941, there were Blenheim IVs operating without undernose guns, like this example wearing the 'RH' codes of No 88 Sqn. After sustaining heavy losses as a Battle squadron in France, No 88 spent much of 1940 and early 1941 'resting', flying coastal patrols from Sydenham, in Northern Ireland. The squadron re-equipped with Bostons fairly soon after its return to East Anglia. This aircraft wears typical Blenheim day-bomber camouflage, with Camotint (Sky) undersides and standard dark earth/dark green topsides. The aircraft is fitted with the later twin Browning mid-upper turret, and has 'Stickleback' flame-damping exhausts.

21
Blenheim I K7059/TW of No 90 Sqn, Bicester, September 1938
Formed as a No 1 Group bomber squadron at RAF Bicester, No 90 switched to training duties with No 6 Group on the outbreak of war, moving from West Raynham to Weston on the Green. The unit eventually merged with No 35 Sqn at Upwood to form No 17 OTU. No 90 Sqn quickly painted its aircraft in warlike colours as the conflict loomed, replacing the giveaway '90' code prefix with a two-letter code. This aircraft already has a Type B red/blue fuselage roundel (achieved by modification, overpainting the white and yellow rings), and unusually has no individual code letter. The 'TW' codes shown here would quickly change to the 'WP' prefix.

22
Blenheim IV N6181/C-SR of No 101 Sqn, West Raynham, August 1940
Like No 90 Sqn, No.101 formed at Bicester and moved to West Raynham, taking a training role when war broke out. But unlike No 90, No 101 Sqn quickly returned to the frontline, flying operational attacks from July 1940, mainly against invasion barges in the channel ports. No 101 had a relatively short life as a Blenheim squadron, re-equipping with Wellingtons in April 1941. The unit 'got around' the limited space for the individual letter between the roundel and the trailing edge by applying it much smaller than the two-letter prefix.

23
Blenheim IV V6028/D-GB of No 105 Sqn, Swanton Morley, July 1941
This aircraft was flown by Wg Cdr Hughie Edwards on the raid on Bremen on 4 July 1941 which won him the Victoria Cross. The individual code letter was applied in a smaller size to fit between the roundel and the trailing edge, and was dropped down to remain clear of the turret, which had the twin-Browning installation. The aircraft had additional firepower in the shape of a free-mounted machine gun in the nose glazing, with another facing aft in the early-style transparent undernose fairing. The aircraft was also fitted with a non-standard rear view mirror above cockpit. Interestingly, when Graham Warner first restored a Bolingbroke to airworthy condition, it was painted in this colour scheme. Sadly, the aircraft crashed on 21 June 1987.

24
Blenheim IV V6374/X-GB of No 105 Sqn, Lossiemouth, late 1941
A handful of the Blenheims used by No 105 Sqn, and some other Bomber Command units engaged in anti-shipping raids, wore the same sea grey and green top-surface colour scheme as was applied to Coastal Command Blenheim IVFs. Thus, this aircraft has the usual Sky undersides, but has dark green and sea grey topsides. The aircraft displays yet another variation in the presentation of its unit codes. Fitted with the two-gun mid-upper turret, the aircraft also has a free-mounted machine gun in the nose and another facing aft in the transparent undernose fairing.

25
Blenheim IV R3816/J-OM of No 107 Sqn, Leuchars, March 1941
While the most obvious (and commonplace) method of preparing No 2 Group Blenheims for night operations was to entirely overpaint their pale Sky undersides with black paint, some aircraft were merely toned down. This No 107 Sqn Blenheim IV has had its pale undersides darkened with a black overspray, while white parts of the national insignia have also been darkened by overpainting with medium sea grey. The toning down of Blenheims began in the winter of 1940, but because night operations played a minor part only a handful of aircraft on a particular squadron would be repainted. No 107 Sqn transferred to Coastal Command for two months in March 1941, flying from Scotland, before returning to Bomber Command in May, and then moving out for a tour of operations on the island of Malta.

26
Blenheim I K7040/V of No 114 Sqn, Wyton, March 1937
This aircraft typifies the initial Blenheim bomber colour scheme, with night undersides, white serials underwing and rough matt dark green/dark earth topsides. On the first 12 aircraft, the serial numbers were

hyphenated, and underwing serials were applied in 30 inch strokes close to the wingtips. Later aircraft had no hyphens, and serials were larger, fatter, and placed further inboard. This aircraft wore A1-type roundels on the fuselage and overwing, and had no fin flash or underwing roundels. The squadron number was applied in 18 inch letters in the flight colour aft of the roundel, sometimes with an individual aircraft letter forward in the same colour, or sometimes with the individual letter limited to a small code on the side of the nose. These original squadron numbers and code letters were soon toned down to a uniform grey. Following the Munich Crisis, the squadron numbers were removed, replaced by two-letter codes (inevitably applied in grey). Some squadrons initially had no codes allocated, and some aircraft flew around without any identifying marks for some weeks. At Wyton, B-type red/blue roundels were applied by over-painting the full-colour A1 roundels. At Bassingbourn, however, A1 roundels were retained. Underwing serials were often removed, and fuselage and fin serials sometimes disappeared too.

27
Blenheim IV N6155/F-FD of No 114 Sqn, Wyton, May 1939

Although designed to meet an Army Co-operation Command requirement for a recce bomber, the Blenheim IV demonstrated such an improvement over the original Mk I that it was rushed into service to replace the earlier version. No 114 Sqn received its first Mk IVs in April 1939, by which time two-letter codes had replaced the original squadron number code. This Blenheim IV is depicted as it appeared soon after its arrival at Wyton, with factory-fresh black undersides, dark green/dark earth topsides, white serials underwing and with no finflash. The unit badge is carried in a 'grenade', but the aircraft already wears a B type fuselage roundel, widely adopted after Munich.

28
Blenheim IV L8756/E-XD of No 139 Sqn, Plivot, France, April 1940

For daylight operations deep into German airspace, black undersurfaces were soon deemed unsuitable, and on 27 October 1939 a Wyton-based Blenheim was sent to Heston, where Titanine repainted it with light sea green (later known as duck egg green) undersides, while its upper surfaces were filled and polished. This gave a useful increase in speed, and the rest of the Wyton-based Blenheim IVs (Nos 114 and 139 Sqns) were soon similarly painted. The first aircraft with the new pale undersides tended to have a high, soft-edged and slightly wavy demarcation between the new Camotint and the original green/earth topsides, while later aircraft had a harder demarcation, sometimes very straight edged, sometimes with tiny scallops. No 139 Sqn's aircraft also tended to have very thin and spindly code letters.

29
Blenheim I L6670/UQ of No 211 Sqn, Menidi/Tatoi, late 1940

Vokes filters and tan/dark earth topsides mark out this

Blenheim I as a Desert Air Force aircraft, although it is portrayed as it appeared during the futile defence of Greece. The aircraft retained only the dark earth of its original scheme, with faded Mediterranean or azure blue undersides, and tan replacing the original dark green on the topsides. The Blenheim has a wide full height fin flash and an A1 fuselage roundel. Although having no individual code letter forward of the roundel, the aircraft does boast the unit's 'UQ' codes in dark red aft.

30
Blenheim IVF R3965/P-LA of No 235 Sqn, Bircham Newton, mid-1940

No 235 Sqn transferred to Coastal Command in February 1940, and began operations in May from Bircham Newton. Initially Coastal Command Blenheim fighters wore standard day fighter camouflage, as shown here, but some eventually received a more suitable two-tone blue-grey and blue-green upper surface camouflage. Operating in the intruder role, the Blenheim IVFs tended to carry light bomb carriers externally, behind their ventral gun packs.

31
Blenheim IF L1336/WR-E of No 248 Sqn, Hendon, late 1939

This No 248 Sqn Blenheim IF wears standard day-fighter colours of the period, with black and white undersides, divided along the centreline. Since most Blenheim IFs had previously been bombers, with black undersurfaces, the addition of white to the underside of the starboard wing and tailplane (and sometimes to the starboard side of the fuselage, too) was easy and straightforward. There were, however, differences in interpretation of the instructions leading to some units keeping black lower engine cowlings to starboard, and others painting them white. No 248 Sqn moved the fuselage roundel well aft on the port side to leave room for the two-letter code forward. The unit was later transferred to Coastal Command, and many of its later Blenheim IVs wore Coastal Temperate Sea colours with extra dark sea grey and dark slate grey camouflaged upper surfaces.

32
Blenheim IVF V5735/D-QY of No 254 Sqn, Aldergrove, July 1941

Although a Coastal Command unit from January 1940, No 254 Sqn's Blenheims still mostly wore standard day bomber colours a year-and-a-half later, with dark green/dark earth topsides. Installation of the fighter-type ventral gun pack precluded fitting a rearward-facing undernose gun turret, and this aircraft retains the single VGO K-Type machine gun in its mid-upper turret. Photos of aircraft wearing 'QY' codes are often misidentified in print as belonging to No 235 Sqn, although the latter unit actually used the code 'LA'.

33
Blenheim 1F (serial unknown) YN-B of No 601 Sqn, Hendon, summer 1939

Immediately before the war a handful of Blenheims still wore squadron markings. This practise was most common among the Auxiliary units, which had in general not been split again and again to form the cadres of new units, and in which unit pride and esprit de corps was at its strongest. Thus, this No 601 Sqn aircraft wears the unit's winged sword badge in red in a red-framed dark blue spearhead frame. In former times, the spearhead frames on British fighters tended to be white, but No 601 had chosen to tone down the device rather than remove it. Apart from this, the aircraft was typical of Blenheim fighters on the outbreak of war. The aircraft had the usual black and white undersides, with the starboard wing and tailplane in white, but retaining a black cowling, elevator and aileron on the starboard side. The B type fuselage roundel was produced by overpainting the white ring with red, and by painting over the outer yellow ring in the normal camouflage colours. The overwing roundels were more crudely modified, and the original colours showed through more. The individual aircraft letter was repeated in white on a flat panel below the pilot's windscreen.

34
Blenheim IV V6027/W-SL of No 13 OTU, Bicester, spring 1941

There are persistent reports that some OTU Blenheims had trainer yellow undersides, but the vast majority wore standard day bomber colours, like this No 13 OTU aircraft. This bomber is typical of OTU machines, with a rectangular fin flash, A1 fuselage roundel and plain grey codes. The aircraft combines the later two-gun mid-upper turret with the early glazed undernose turret. Hard-edged straight camouflage demarcation is typical of late-production Mk IVs.

35
Blenheim IF K7159/YX-N of No 54 OTU, Church Fenton, September 1941

Often mis-identified in photo captions as an aircraft of No 614 Sqn, K7159 actually served with No 54 OTU, a nightfighter training unit. The aircraft has the standard underfuselage gun pack, and is also fitted with an AI Mk III radar antenna in the nose. It wears a weathered special night finish, with a narrow full height fin flash (colours of equal widths), and an A1 fuselage roundel. The 'YX' codes were applied in medium sea grey forward of the roundel, with the individual letter 'N' aft of the roundel outlined in yellow, making it slightly larger and fatter. The yellow outlined individual code indicated assignment to an OTU, following a directive to all Fighter Command OTUs on 25 May 1941. The serial was applied in tiny characters on the rear fuselage, and was not repeated on the fin or rudder. Most of the lower nose-glazing was over-painted to cut down reflections, since there was no need for visual bombaiming. Overall black proved unsuccessful as a night camouflage, proving too stark and effectively 'silhouetting' the aircraft, and the scheme was short-lived.

36
Blenheim V BA612 of No 132 OTU, East Fortune, mid-1943

No 132 OTU was a conversion training unit whose primary role was to train aircrew destined for Coastal Command Beaufighter squadrons. With no dual-control Beaufighter, the Blenheim V was a natural choice for initial conversion phases of the training syllabus. This aircraft wears standard late-war Coastal Command temperate camouflage, with white undersides and a very high demarcation line separating the white from the medium sea grey topsides. The aircraft wore a C-type fin flash and a C1 fuselage roundel. Since the Blenheim Vs of No 132 OTU were used for pilot training, all gun turrets were removed and faired over.

37
Blenheim IVF Z7513/B of No 15 Sqn, SAAF, Cyrenaica, Libya, April 1942

SAAF Blenheims were supplied by the RAF, and wore RAF serials and standard camouflage schemes. Thus, this aircraft wore the same Azure undersides and earth/mid stone topsides as did Desert Air Force Blenheims assigned to RAF squadrons. The national insignia was also standard RAF style, with a rectangular fin flash with equal stripes, although the A1 fuselage roundel normally had a South African orange centre. No codes were carried, with the individual aircraft letter applied in medium sea grey aft of the roundel, and with the serial in black on the rear fuselage. The aircraft was fitted with a two-gun mid-upper turret, but the ventral gun pack ruled out an undernose turret. The emphasis on ground-strafing was reflected by the provision of a single fixed 20 mm Hispano projecting from the starboard bomb aimer's window.

38
Blenheim V BA328/R of No 13 Sqn, Royal Hellenic Air Force, Aden 1943

Blenheims delivered to Greece before the German invasion were captured when the country fell. But No 13 Sqn (which fled to Egypt with five of its Ansons) was soon re-equipped with Blenheims by the RAF. Initially the squadron used desert-camouflaged Blenheim IVs, but these were transferred to RAF units in Burma in mid-1942 and replaced by ex-RAF Mk Vs. These aircraft retained RAF colours, serials and markings, some in desert colours, others in maritime schemes. This aircraft had white undersides, with extra dark sea grey/dark slate grey topsides with a very high, scalloped demarcation. The aircraft wore a standard C-Type rectangular fin flash, and a C1 fuselage roundel. The code letter 'R' was applied in light slate grey A aft of the roundel with the light slate grey serial further aft on the rear fuselage. The aircraft had a standard Mk V mid-upper turret and undernose turret fairing, but with no guns in the latter position.

39
Blenheim IV of *Groupe Lorraine*, Free French Air Force, attached to No 270 Wing, RAF, Fuka, Western Desert (Egypt), November 1941

Free French Blenheim IVs tended to wear standard RAF-type desert camouflage, with azure undersides, and earth/mid stone topsides. But markings varied enormously, usually with French-type roundels and rudder striping, with blue leading, and with the red ring of the roundel outermost. Some aircraft carried the Free French red cross of Lorraine simply outlined in white, aft of the roundel, while others had the cross of Lorraine on a white disc. This sometimes augmented or even replaced the fuselage roundel, and was often applied inboard of the wing roundels. Some aircraft had an RAF-style full-height type fin flash instead of the rudder stripes, but in French colours, with blue leading. RAF serials were usually retained on the rear fuselage, and two-digit numerical codes were often applied to the tailfin. Some aircraft carried the insignia of *Groupe Lorraine*, which consisted of a white shield with three spreadeagled bird shapes superimposed on a red diagonal.

40
Blenheim V BA849 of *Groupe Bretagne*, Free French Air Force, initially attached to the RAF's No 203 Sqn, Ben Gardane, Tunisia, April 1943

This Blenheim V wore unusual colours, with prominent black and white bands encircling the outer wing panels immediately outboard of the engines and around the fuselage, behind the wing trailing edge. These were applied over the standard desert colour scheme of azure undersides, with earth and mid stone topsides. No codes or roundel were carried on the fuselage, but national insignia was applied in the from of French-type rudder striping, and with French roundels on the wings. The RAF serial was carried on the rear fuselage in black. The aircraft had no mid-upper turret, instead being fitted with a fairing incorporating a small astrodome. Similarly, the Blenheim had no undernose gun fairing, but did have a blister above the nose glazing's middle 'roof' panel. This aircraft crashed in the sea off Tunisia on 23 April 1943.

FIGURE PLATES

1
Sgt George Keel, Wireless Operator/Air Gunner with No 235 Sqn at Thorney Island in October 1940. He is wearing a 1930 Pattern flying suit with fur collars, beneath which he would have on his standard battle dress. Keel's large leather gauntlets are 1933 Pattern Issue, whilst his helmet is a Type B and the goggles poking out of the left thigh pocket are Mk IIIA specification. Finally, his leather flying flying bootsare 1936 Pattern Issue. Twenty-year-old George Keel was posted missing, along with the two remaining members of his crew, following an engagement between Blenheim IV N3530/'QY-S' and enemy aircraft during an airfield protection patrol south of Thorney Island in the late afternoon of 9 October 1940.

2
Sgt S W Lee, wireless Operator/Air Gunner with No 113 Sqn at Heliopolis, in Egypt, in January 1940 is adorned in a full two-piece fur-lined Irvin suit. His 1930 Pattern boots are a matching leather brown, as is his Type B quartered helmet – a Type D full-face mask is secured to the latter. Lee is wearing a full parachute harness, with the pack slung snugly beneath his bottom. A veteran a much action in the Mediterranean, Lee was eventually promoted to flight lieutenant, and awarded the AFC and DFM.

3
Sgt Stuart Bastin, Wireless Operator/Air Gunner with No 105 Sqn at Swanton Morley in July 1941. Bastin is wearing the standard issue No 1 uniform, complete with shiny steel-capped shoes. Over his Mae West he has the wide straps of the seat-type parachute harness. Bastin was lost with his crew on 26 August 1941 when Blenheim IV Z7682/'N' hit the mast of a ship it was attacking off the Tunisian coast and exploded.

4
Sqn Ldr Hector Lawrence, Officer Commanding B Flight of No XV Sqn at Wyton in May 1940, is wearing a Flying Suit, Combined Pattern in blue cotton twill (note his rank stripes on the suit cuffs), beneath which is worn standard battle dress. He has slightly modified the suit by removing the connectors for the chest-type parachute, leaving the connectors for the seat-type parachute only. Lawrence's footwear comprised a pair of 1936 Pattern flying boots, kept in immaculate condition. A veteran of the Albert Canal raid on 12 May, Lawrence was killed in action six days later when his formation was bounced by Bf 109Es whilst attempting to bomb columns of Wehrmacht troops advancing on Le Cateaux.

5
Sgt T Inman, pilot with No 82 Sqn at Watton in March 1941, is wearing An Irvin 'Harnisuit', specially developed for bomber crews, over his No 1 uniform. It had inflatable panels and three attachment points for various types of parachute. His helmet is a Type B and the oxygen mask a Type D.

6
Regular NCO Sgt J W Davies, navigator with No 600 Sqn at Manston in June 1940, arrived back from Holland dressed in this attire after having been 'on the run' from the Germans since being shot down on 10 May in Blenheim IF L6616/'BQ-R' during an attack on Waalhaven. Davies was killed (along with his pilot, Sgt A F C Saunders) in a flying accident in L6684 on 7 September 1940, his aircraft suffering port engine failure during a landing approach to Hornchurch and crashing on its back from 200ft.